# Master the Art of Soldering with this Must-Have Guide

Liam N. Watson

*Unlock the Secrets to Perfecting Soldering Skills with Expert Guidance and Foolproof Methods*

## Life advices:

*Stay proactive in health check-ups; prevention is better than cure.*

*Your actions echo in eternity; make choices that reflect the legacy you wish to leave.*

# Introduction

This book is a comprehensive resource that provides essential information and step-by-step instructions for soldering. Whether you're a beginner or looking to improve your soldering skills, this guide covers all the basics you need to know.

The guide starts by discussing the different choices of soldering irons and their features. It explains the importance of selecting the right soldering iron for your specific soldering tasks and provides insights into temperature control and wattage considerations.

Next, the guide dives into the process of soldering. It covers important topics such as cleanliness, tinning the soldering iron tip, and preparing for soldering. It explains the types of solder and fluxes used in soldering and provides guidance on their proper use. The guide also highlights the importance of temperature control and provides tips for achieving proper solder flow.

Step-by-step instructions are provided to guide you through the soldering process. It covers various soldering techniques, including interwiring and wire joints, and provides tips for tidying up your soldered connections. The guide also introduces the reflow technique, which is useful for repairing and reworking soldered components.

To address common issues that may arise during soldering, the guide includes sections on fatigue and breakage, as well as faults and desoldering techniques. It explains how to use desolder braid effectively for removing solder and provides troubleshooting tips for identifying and resolving soldering problems.

The guide also includes a quick summary guide for easy reference, providing a concise overview of the soldering process and key considerations. Additionally, it offers a troubleshooting guide to help you diagnose and fix common soldering issues.

Safety is a crucial aspect of soldering, and the guide addresses possible hazards and provides simple first aid tips in case of accidents or injuries.

This is a valuable resource for anyone interested in soldering, from beginners to experienced enthusiasts. By following the guidance and instructions in this guide, you can develop the skills and knowledge needed to create reliable and professional soldered connections for various electronic projects.

# Contents

# First steps

The principle behind soldering sounds quite simple: the idea is to join components together to form an electrical connection, by using a mixture of lead and tin solder or alternatively "lead-free" solder (an alloy of tin and copper), which is melted onto the joint using a *soldering iron*. If you have never picked up a soldering iron before, then this guide will show you everything to help you start soldering with confidence. I also hope that the guidance will help those working in other areas — computer technicians or audio enthusiasts, for example, who may be faced with electronic repairs or modifications using a soldering iron for the first time.

If you're an electronics hobbyist or trainee, before embarking on any form of ambitious electronic project, it is recommended that you practice your soldering technique on some *brand new* components using *clean* strip board (or protoboard) or a printed circuit board, and select a simple and straightforward constructional design as a starting point. Become acquainted and comfortable with your chosen soldering iron, which likely to become as familiar to you as a favourite pen. Learn how to balance it and handle it with precision. Try soldering an assortment of resistors, capacitors, diodes, transistors and integrated circuits with it, and then try your hand at *desoldering* – removing the solder again to make a repair or modification.

A really good place to start learning is by building a simple electronics kit, such as a Velleman kit that contains a good quality printed circuit board. You'll learn some of the basic skills of successful soldering this way, and it'll be a great confidence booster too.

Did you know? In the USA and elsewhere, the letter L in "solder" is silent and they say "soda" or "sodder" – but here in Britain we do pronounce the L and we say "sole-der".

# Soldering iron choices

Search any electronics catalogue or website and you'll see a bewildering array of soldering equipment on sale, including irons, controllers, work stations and desoldering equipment too. A large range of soldering irons is readily obtainable - which one is suitable for you depends on your budget and how serious your interest in electronics is, but there's something for every pocket distributed by a variety of retail, industrial and mail-order outlets.

The Antex range of soldering equipment has been very popular with industry, education and the electronics hobbyist for 60 years; I grew up with an Antex iron and a trusty 15W Antex iron was an everyday part of my hobby electronics all the way through the 1980's. An industrial user or a more dedicated hobbyist – with a bigger budget! – will be interested in a soldering station instead and again Antex offers a range of British-made products for industry or home use.

A very basic mains electric soldering iron can cost from under £5 (US$ 8), but I find that these very cheap irons, as sold on auction websites, are pretty crude and imprecise. They are best suited for simple electrical repairs and DIY rather than precision electronics or printed circuit boards discussed here. They tend to be bulky and uncomfortable for extended use, and they may not have suitable "bits" or tips of various sizes to suit different tasks.

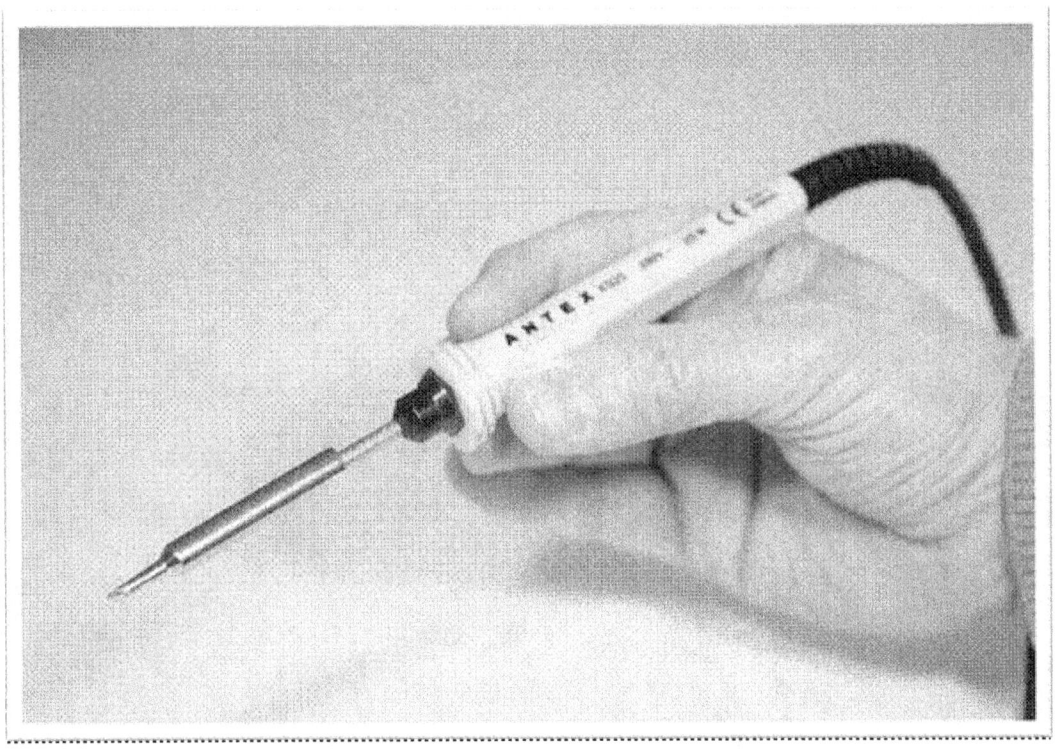

*⌃⌃ This classic Antex XS25 "pencil style" 25W mains-powered soldering iron has exchangeable tips or "bits" and is great for general hobbyist or educational use. Stands are also available to store them safely in between use.*

A quality pencil-style electric soldering iron such as the Antex XS25 (photo) will be approximately £20 (US$18 tax free) - though it's possible to spend into three figures on a soldering iron "station" if you're really serious about the subject! Don't be tempted to over-spend on an elaborate workstation though, unless you are really very serious about becoming involved in electronics. You will usually obtain perfectly satisfactory results using a fairly modest "pencil" iron model, and you can upgrade to something more sophisticated should your needs change in the future.

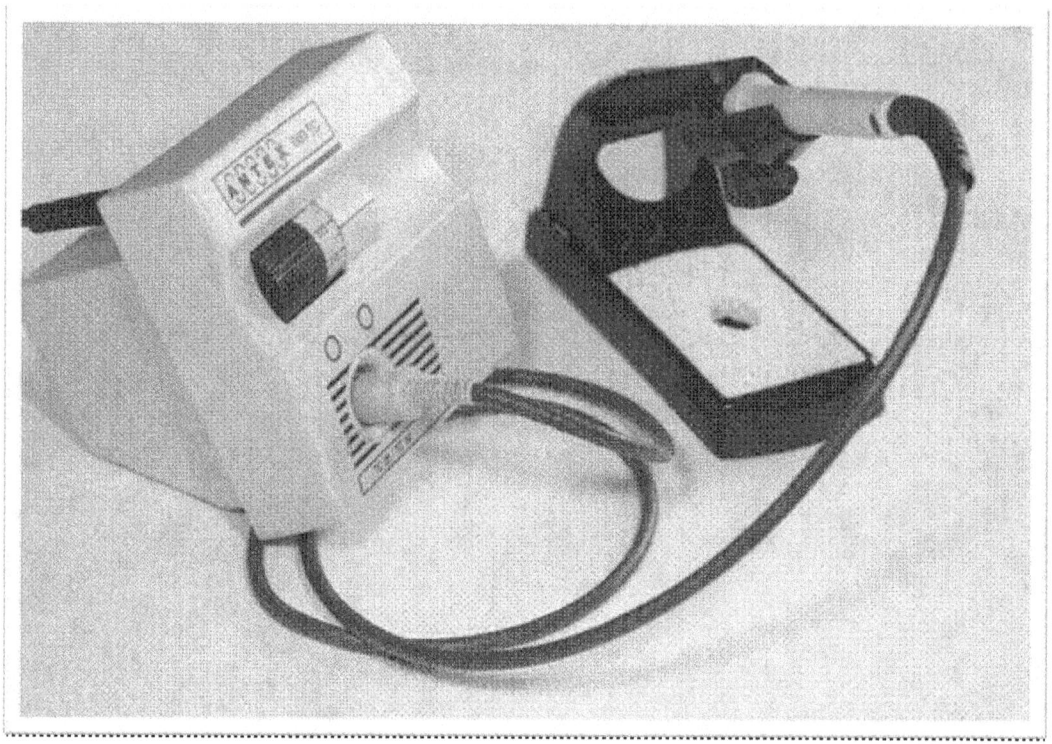

^^ *For enthusiasts or industry, this Antex 660TC soldering station has a separate mains-powered control unit and a matching low-voltage soldering iron rated at 24 Volts, 50 Watts so it's suited to a wider range of tasks than a lower powered one.*

When choosing your soldering iron, certain factors to bear in mind include:

**Voltage**: for the British market, "mains" electric irons run directly from the mains at 230V a.c. or will be set for other voltages (110V a.c.) depending on the country. However, **low voltage types** (e.g. 12V or 24V) usually form part of a "soldering station" for use with a matching controller made by the same manufacturer. Some low-voltage irons run off batteries (e.g. a car battery or Ni-cads) but these are uncommon.

**Wattage**: this is an extremely important factor to think about when choosing your iron. Typically, irons for general electronics work may have a power rating of between 15-25 watts or so, which is fine for most electronic assembly tasks, printed circuit boards and inter-wiring. It's important to note that a higher wattage does *not* mean that the iron runs hotter - it simply means that there is more power "in reserve" for coping with larger joints. The maximum electric iron wattage generally available is about 100W, which is OK for DIY electrical repairs but is far too high for general electronics or circuit board use.

A higher wattage iron offers you more flexibility for tackling a wider range of tasks. It has a better "recovery rate" which makes it more "unstoppable" when it comes to heavier-duty work, because it won't be drained of its heat so quickly. So check the power ratings carefully, and anything between 15–40W is fine for general electronics soldering.

**Temperature Control** - the simplest and cheapest types don't have any form of temperature "regulation". Simply plug them into the mains and switch them on! Thermal regulation is "designed in" (by physics, not electronics!). Sometimes they are described as "thermally balanced" as they have some degree of temperature "matching" – in other words, they warm up as quickly as they lose heat during use, so in a primitive way they maintain roughly a constant temperature. This type of iron is perfectly acceptable for hobby or less demanding professional use. It's essential to use the manufacturer's specified tips (see later) to maintain proper temperature matching, otherwise the iron may not heat up enough – or it may overshoot in temperature.

These unregulated irons form an ideal general-purpose iron for most users, and they cope reasonably well with printed circuit board soldering and general interwiring. However, most of these "miniature" types of iron will be of little use when attempting to solder large joints (e.g. very large terminals or very thick copper wires) because the components being soldered will draw or "sink" heat away from the tip of the iron, cooling it down too much and preventing solder from flowing properly. That's when a higher wattage iron is needed.

A proper **temperature-controlled iron** will be quite a lot more expensive - retailing at say £40 (US$ 60) or more - and will have some form of built-in thermostatic control, to ensure that the temperature of the "bit" (the tip of the iron) is maintained at a fixed level within reasonable limits. This is desirable during frequent use, since it helps to ensure that the temperature will be relatively stable regardless of the workload. Some irons have a bimetallic strip thermostat built into the handle which gives an audible "click" in use, and some may include an adjustable screwdriver control within the handle as well. Others may have electronic controls built in.

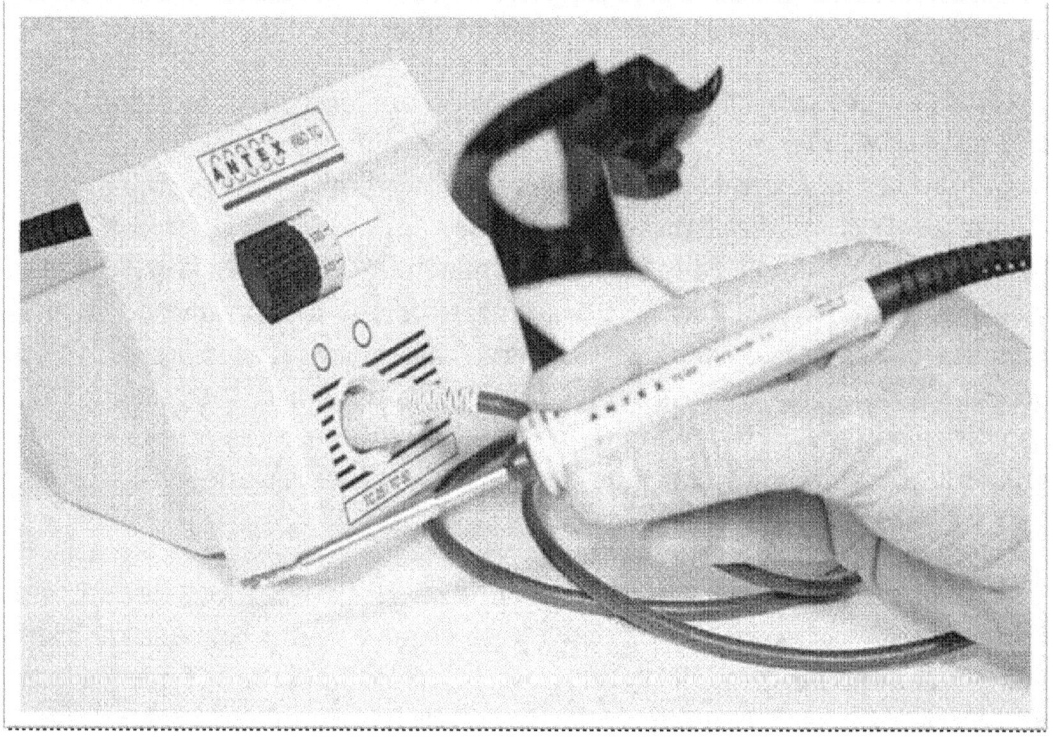

*^^ An Antex 660TC soldering station with matching iron and bench stand.*

More expensive still, **soldering stations** cost from £70 (US$ 115) upwards (the iron may be sold separately, so you can pick the type you prefer), and consist of a complete bench-top control unit into which a special *low-voltage* soldering iron is plugged. Some versions might have a built-in digital temperature readout, and a control knob to vary the setting. The temperature could be boosted for soldering larger joints, for example, or for using higher melting-point solders (e.g. lead-free or silver solder). These are designed for the discerning user, or for continuous production line or professional use. A thermocouple will be built into the tip or shaft, which monitors temperature.

The best soldering stations have irons which are well balanced, with comfort-grip handles which remain cool all day, and silicone-based cables which are burn proof. Antex produces a range of irons with silicone cables specially for education use, to help avoid accidents caused by careless use by students.

**Anti-static protection**: if you need to solder a lot of static-sensitive parts (e.g. CMOS chips or MOSFET transistors), more advanced and expensive soldering iron stations use static-dissipative materials in their construction to avoid static charges accumulating on the iron itself, which could otherwise damage or "zap" some semiconductors. Such irons are listed as "ESD safe" (electro-static discharge proof). The cheapest irons are not classed as ESD-safe but they still perform well enough in most hobby or educational applications provided you take the usual anti-static precautions when handling these components. The iron would need to be well earthed (grounded) in these circumstances, to carry away any static.

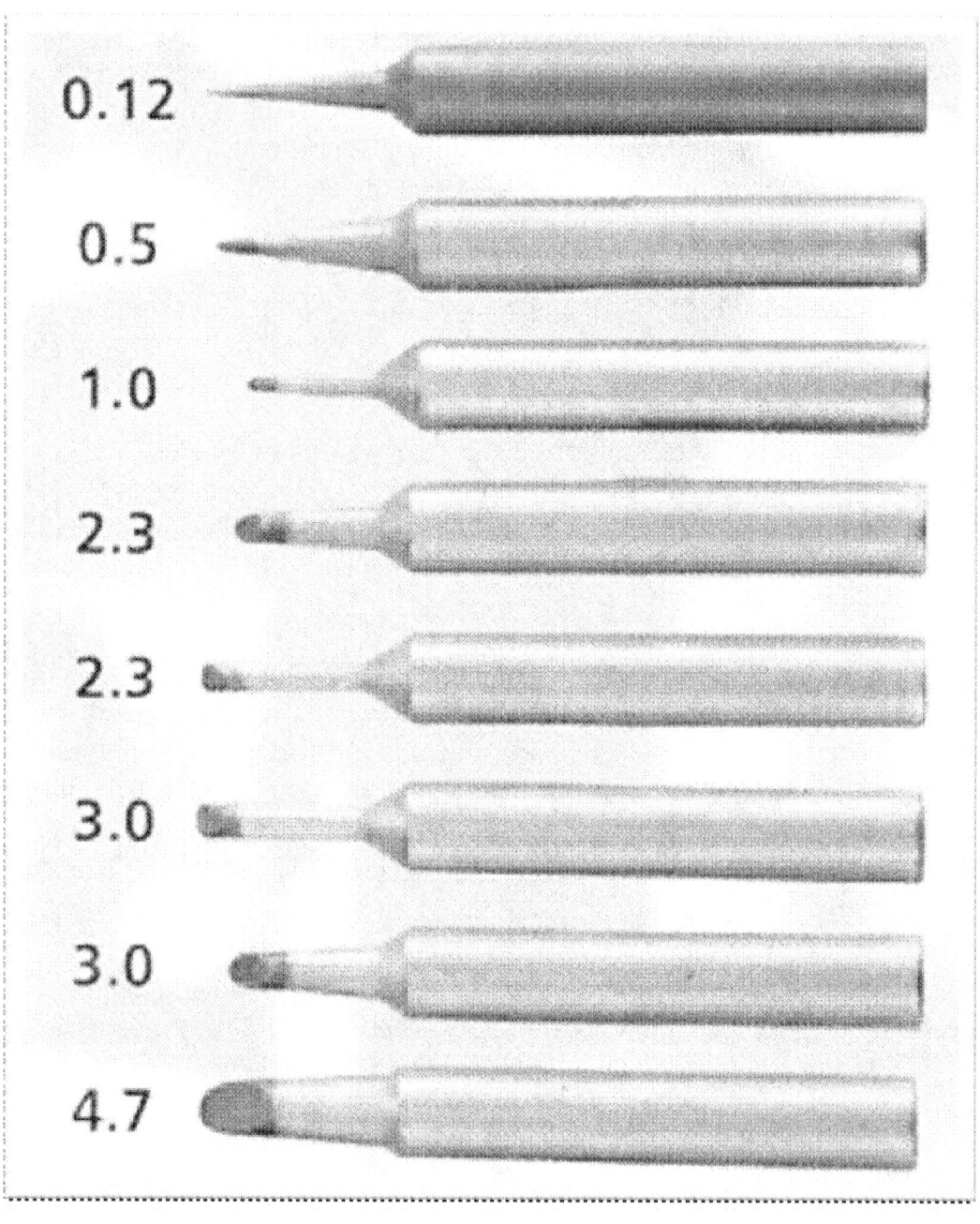

0.12

0.5

1.0

2.3

2.3

3.0

3.0

4.7

^^ *Just part of the range of spare tips or "bits" produced by Antex*
*for their soldering irons. Use only tips intended for a particular iron, to*

*avoid thermal matching problems.*

**Tips or Bits**: it's often useful to have a small selection of manufacturer's bits (soldering iron tips) available with different diameters or shapes, which can be changed depending on the type of work in hand. You will probably find that you become accustomed to, and work best with, one particular shape of tip for the majority of your work. Usually, tips are iron-coated or plated to preserve their life and to maintain good tip "hygiene". Be sure only to use tips that are specifically designed for your iron, otherwise thermal problems may arise. I show separately some typical tips, courtesy of Antex.

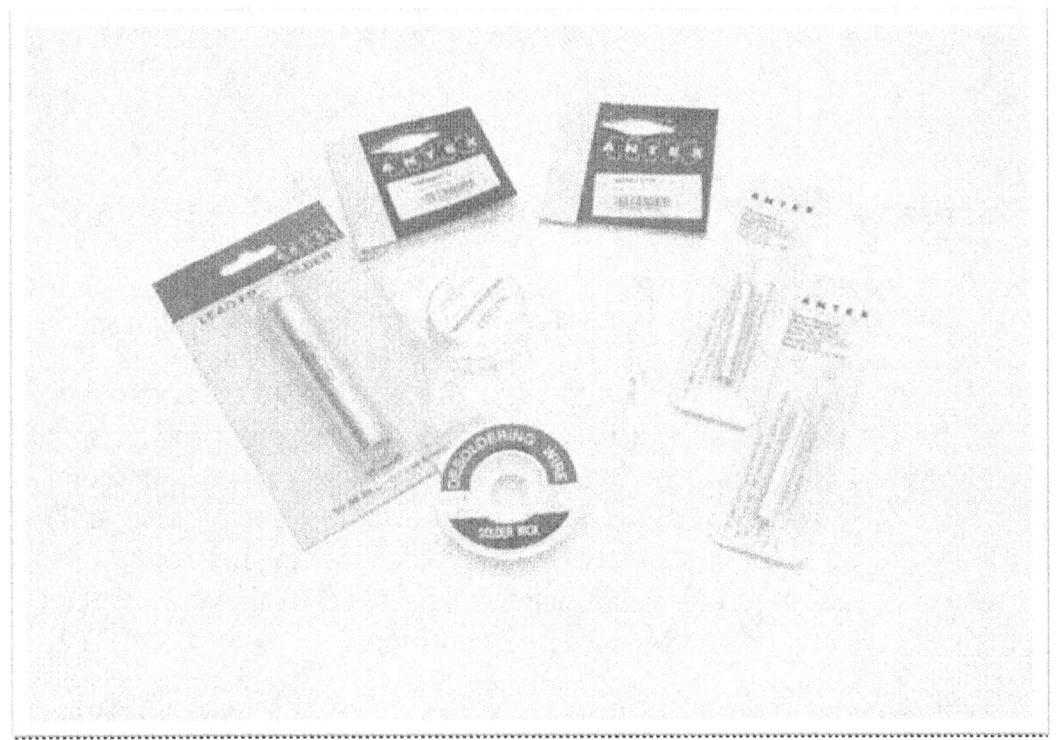

*^^ Useful accessories from Antex including solder, tips and tip cleaner, desoldering braid and a heatshunt.*

**Spare parts:** it's always reassuring to know that spares are available in the future if required, so if the element blows, you don't need to replace the entire iron. This is especially the case with expensive irons. Check some websites or mail-order catalogues to see whether spare parts are listed. One drawback is that you may need another soldering iron when exchanging a broken heating element!

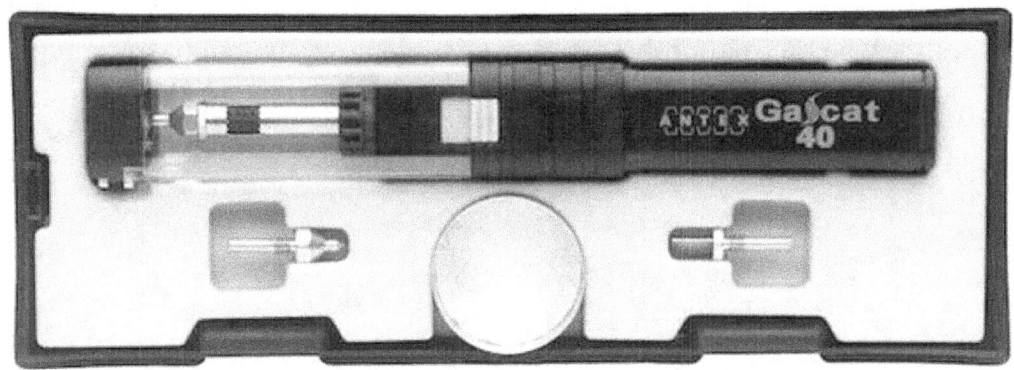

^^ *Gascat 40 gas-powered soldering iron kit by Antex, with spare tips and cleaning sponge.*

**Gas or electric?** So far I've discussed electric soldering irons, but **gas-powered** soldering irons are sold which use butane propellant rather than mains electricity to operate. They have a flint lighter or (better still) a built-in piezo for ignition, and have a catalytic element which, once warmed up, continues to glow hot when gas passes over them. They tend to be big and bulky compared to electric pencil irons.

Field service engineers use gas-powered irons for working on repairs where there may be no power available, or where a joint is tricky to reach with a normal electric iron, so they are really for occasional "on the spot" use, rather than for mainstream construction or assembly work. I use one when I just need to do a quick repair and can't be bothered getting the electric soldering iron going!

Gas irons can have higher power equivalents than electric ones (eg 125 watts or more) but some gas-powered irons are nothing more than miniaturised blowtorches, which may or may not be useful for occasional heavier duty soldering. In the author's experience they can be difficult to use in confined areas. *Extreme care is needed at all times to ensure hot gas emitted from the iron's exhaust port doesn't damage nearby materials, plastics or wiring.* Gas irons can have useful accessories to convert them into e.g. a hot knife for sealing nylon rope, or a hot-air blower for use with heatshrink tubing. Almost every electronics constructor uses an electric-powered iron though.

A **solder gun** is a pistol-shaped iron, typically running at 100W or more, and is completely unsuitable for soldering modern electronic components: they're too hot, heavy and unwieldy for micro-electronics use, nor are they designed for that. Plumbing or DIY, maybe..!

*^^ A heat resistant soldering iron stand with cleaning sponge. (Antex)*

Soldering irons are best used along with a **heat-resistant bench stand,** where the hot iron can be safely stored in between use (photo). It is extremely important that a hot soldering iron is safely "parked" ready for action, and a bench stand is really a necessity. Soldering stations usually have such a feature, otherwise a separate soldering iron stand is essential, ideally one that's supplied with a tip-cleaning sponge. You can make your own cleaning sponges using *cellulose* sponge only.

^^ *A benchtop fume extractor fan for hobbyist use. A replaceable carbon filter helps remove particles and air is vented out the back.*

Other equipment worth considering includes the use of **fume extractors**, which are compulsory in the industrial workplace. Solder fumes and flux smoke are not known to be toxic but they can cause irritation. A basic fume extractor (photo) consists of a small bench-top fan which draw fumes and irritating smoke away from the operator's face and filters out some of the smoke particles, before exhausting the air back out through the fan. The carbon-impregnated foam filters are replaceable. Such devices are quite effective and users soon find them indispensable, even though they can be a bit noisy at close range.

Professional fume extraction systems draw the smoke and fumes directly from the work area via a clip-on tube fitted to the soldering iron, then vent the fumes away through a large filter pump. It is definitely worth considering a small bench top unit for regular hobby or occasional professional use, as having decent ventilation can only be a good thing.

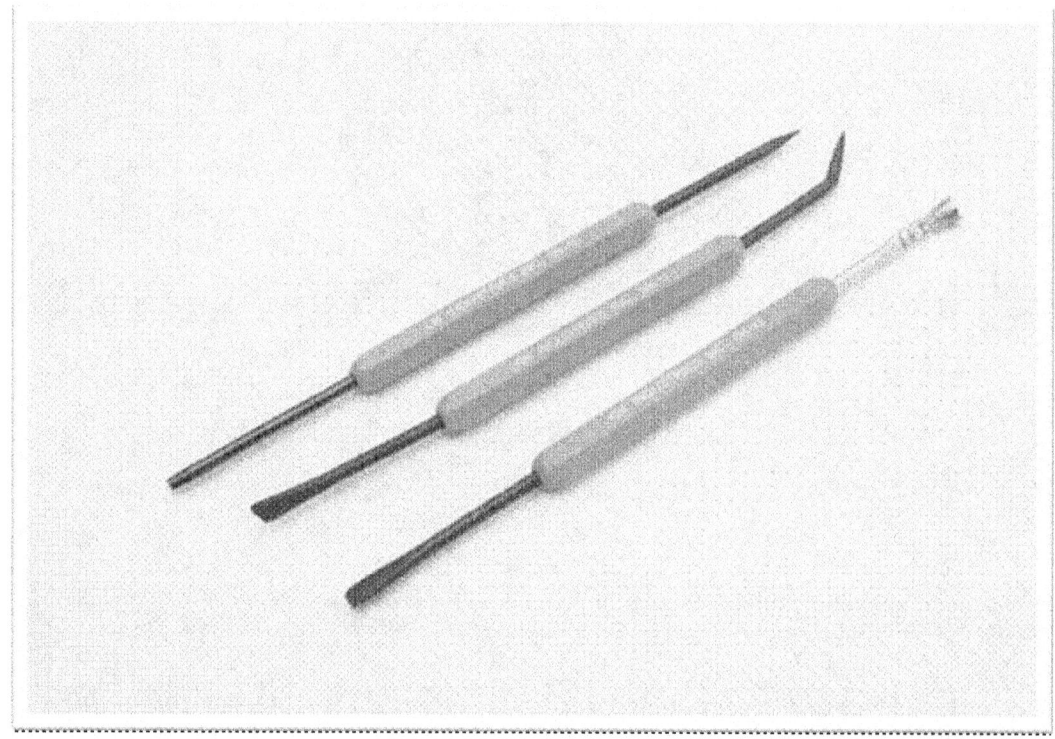

^^ *A soldering accessory toolkit by OK Industries, including probes, scrapers and wire brush*

A variety of specialist **hand tools** are available that assist with soldering, and a good supplier's catalogue will offer a range of small brushes, scrapers and cleaning tools in a handy kit, together with the usual types of wire cutters, pliers and so forth, which are necessary for handling components and tidying up as required. Some specialist service aids (aerosols etc.) are also described later.

Now let's look at how to use soldering irons properly, and later on we will describe the techniques for putting things right when a joint somehow goes wrong — and don't worry, even the experts get it wrong sometimes!

## How to solder

This guide will show you step by step how to solder successfully and plenty of photographs are provided to help explain the techniques needed.As you read through the guide I'll explain all the stages in more detail, but let's look at the basics first.

First of all, successful soldering requires that the items being soldered together are held with as little movement as possible. So it's best to *secure the work* as needed, so that your accuracy isn't affected should the workpiece move accidentally.

In the case of a printed circuit board, various holding frames are useful especially when densely populated boards are being soldered: the idea is to insert all the parts on one side (a process called "stuffing the board"), hold them in place with a suitable pad to prevent them falling out again, turn the board over and then snip off the wires with cutters before soldering the joints. The frame saves an awful lot of turning the board over and back again, especially with large boards: all the soldering can be performed in one pass.

*^^ The ever-popular "Helping Hands" (left) helps support sundry parts, wires etc. during soldering.*
*A modeller's vice (right) holds parts firmly. A vacuum base fixes it onto smooth surfaces.*

Only the more serious constructor will purchase a holding frame, and hobbyists can retain parts in place by improvising in a variety of ways – the ever popular "Helping Hands" stands cost a few pounds and is widely sold. They have adjustable crocodile clips to grip parts, and maybe a magnifying glass or soldering iron stand as well. The cast iron base provides stability. Other parts could be held firm in a modeller's small vice, for example.

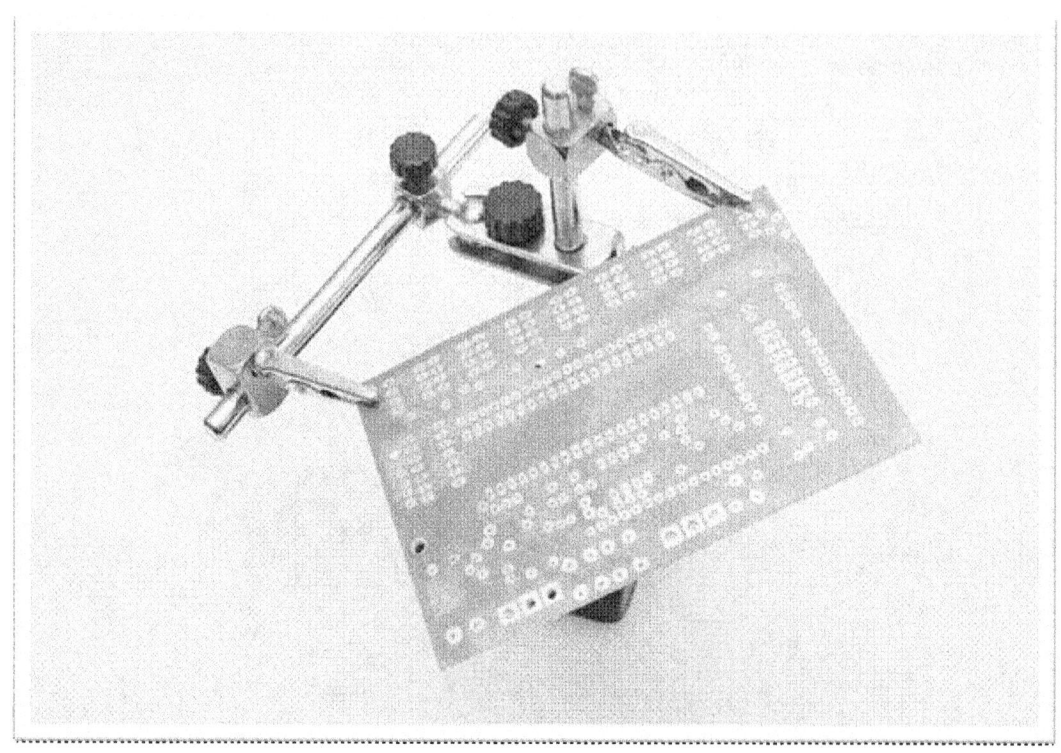

^^ *"Helping Hands" uses crocodile clips to grip parts during soldering. Or just place parts flat on the bench.*

When soldering parts onto an ordinary circuit board, components' wires can simply be bent to the correct pitch (distance apart) to fit through the board, insert the part flush down against the board's surface, splay the wires a little so that the component grips the board under spring tension, and then solder it. This technique is universally used in manual **through-hole soldering,** which is explained in full later.

In the author's view - opinions vary – it's generally better to snip off the surplus wires leads first, to make the joint and neighbouring joints more accessible and also to reduce the mechanical shock transmitted to the p.c.b. copper foil. However, in the case of diodes and transistors the author tends to leave the snipping until after the joint has been made, since the excess wire will help to "sink" heat away from the heat-sensitive semiconductor junction.

A special clip-on heatsink is available which also helps stop excess heat from reaching temperature-sensitive semiconductors like these. I've always managed without one but beginners might find them reassuring. Integrated circuits can either be soldered directly into place if you are confident enough, or better, use a dual-in-line socket to prevent heat damage. The chip can then be swapped out at a later date if needed.

Parts which become hot in operation (e.g. some power resistors), should be raised above the board slightly to allow air to circulate. Some components, especially large electrolytic capacitors, may require a mounting clip to be screwed down to the board first, otherwise the part may eventually break off due to vibration. It's a good idea to bolt such components firmly into place before soldering their terminals, in order not to strain the soldered joints or the components when fasteners are tightened.

By securing the workpiece as much as possible to prevent movement, you have a much higher chance of producing good-quality reliable solder joints.

## Next steps

Let's get to grips with actual soldering techniques in more detail. The soldering of electronics components utilises **lead/ tin** or **lead-free solder** and the process is compatible with many non-ferrous metals. You can solder copper, lead, brass, gold plate, silver, nickel, tin and tin plate, zinc and more besides but some metals such as nichrome, galvanised or stainless steel require a highly specialist "flux" (see later) to solder them and aren't discussed here.  Some materials such as beryllium, chromium, magnesium and titanium are non-solderable in any case, according to solder manufacturers Multicore.

In electronics we're mainly concerned with soldering parts or wires onto printed circuit boards or terminals that are usually already "tinned" with solder or plated, ready for soldering with flux-cored solder. The key factors affecting the quality of a solder joint are:

- **Cleanliness** – dirt or impurities drastically hinder good solder coverage.
- **Temperature** – the right level to enable the solder to flow freely!
- **Time** – apply heat for just the right amount of time!
- **Adequate solder coverage** – enough to form a good joint without touching neighbouring areas.

A little effort spent now in soldering the perfect joint may save you — or somebody else —a considerable amount of time in troubleshooting a defective joint in the future. Let's discuss the basic principles outlined above in more depth.

## Cleanliness and "Tinning" the bit

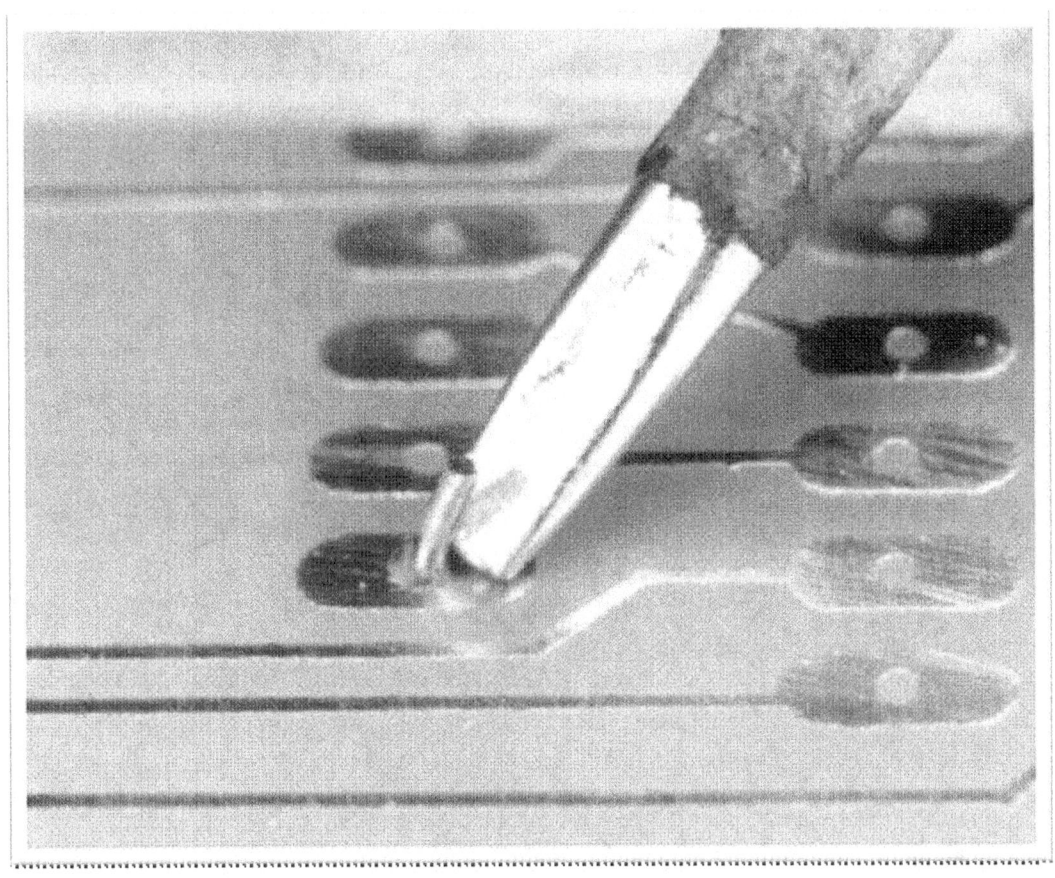

*^^ A clean and shiny soldering tip or "bit" is essential for each and every solder joint you make.*

Firstly, and without exception, all parts - including the soldering iron's tip itself - must be clean and free from contamination. Fact is, solder just will not "take" to dirty parts! Old components or copper board can be impossible to solder because of oxidation that builds up on the surface of the leads. This repels the molten solder and will soon become evident because the solder will "bead" into shiny globules looking like mercury, going everywhere except where you need it. *Dirt and contamination are the enemies of a good quality soldered joint!*

When all the conditions are right for soldering, materials are said to be "wettable" and will accept molten solder, which should flow readily over their surfaces. Hence, it's **really necessary** to ensure that parts are free from grease, oxidation and other contaminants. Note that in any case, some incompatible materials or surface finishes just cannot be soldered using ordinary lead-free solder, no matter how hard you try – e.g. aluminium parts would require special aluminium solder and fluxes (see later) to be used.

*^^ The two tags on this switched potentiometer have blackened with age and oxidation. They must be cleaned before they can be soldered...*

In the case of old components that have been stored a long time, where the leads have blackened and oxidised, use a small hand-held file or perhaps fine emery cloth to reveal fresh metal underneath. Stripboard and copper printed circuit board will generally oxidise after a few months, especially where it has tarnished due to fingerprints, so the copper strips could be cleaned using e.g. an abrasive rubber block made for the purpose. Also available is a fibre-glass filament brush (photo), which is used propelling-pencil-like to remove contamination. They're OK for general cleaning but best avoided on fine surfaces (e.g. gold plated switch contacts). These produce tiny particles which can irritate the skin, so avoid accidental contact with any debris.

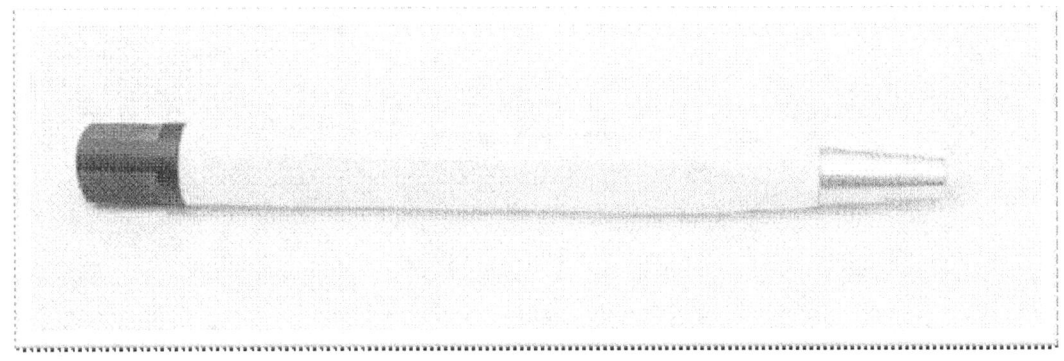

*^^ A glass-fibre filament  brush like this is useful for cleaning oxidised parts.*

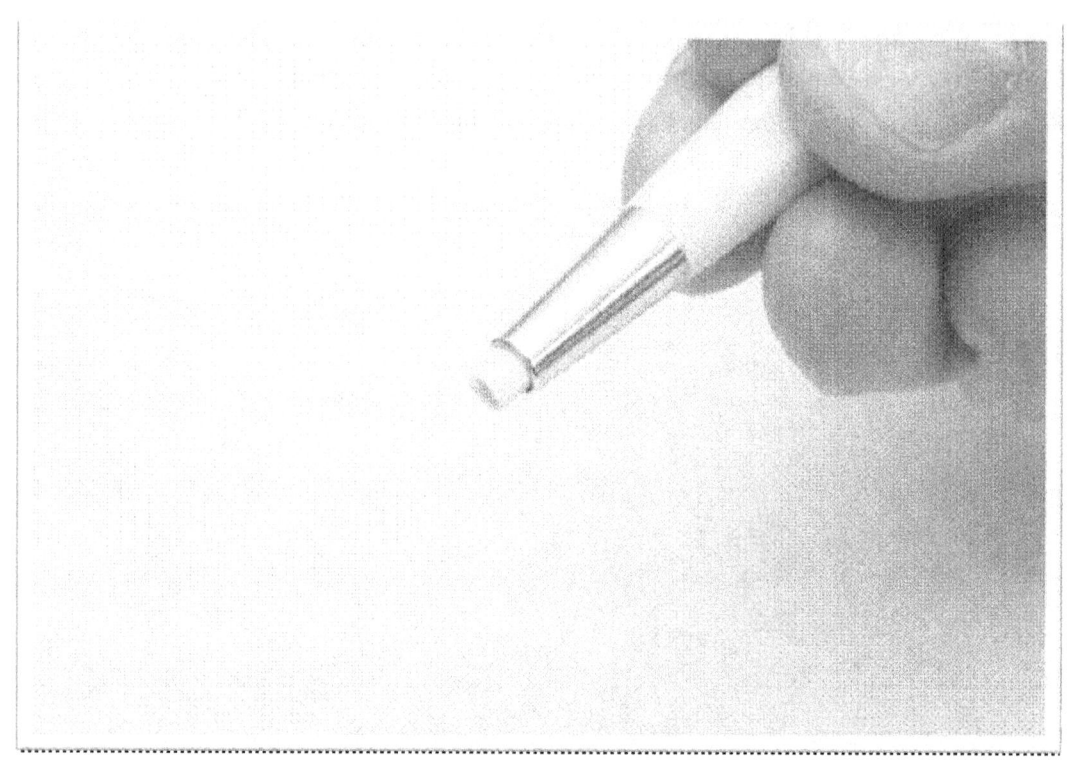

*^^ The glass-fibre brush works like a propelling pencil and produces irritating dust.*

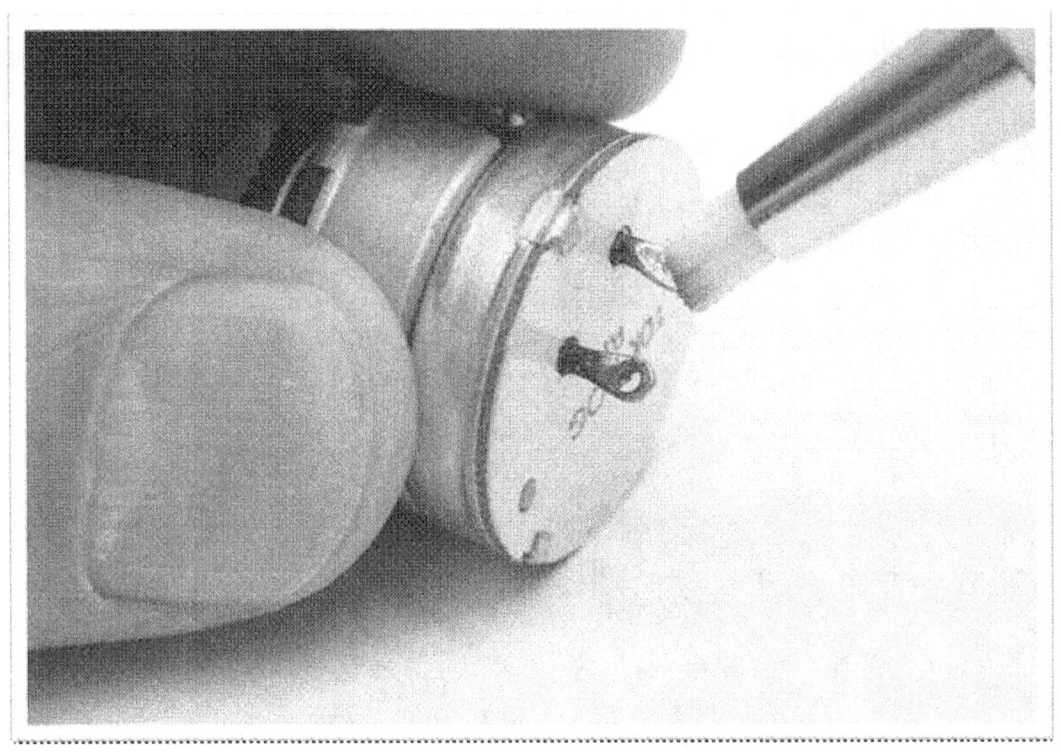

*^^ Cleaning oxidised tags with a glass fibre brush to make them nice and shiny, ready for soldering*

Afterwards, a wipe with a rag soaked in cleaning solvent will remove most grease marks and fingerprints. After preparing the surfaces, avoid touching any parts as far as possible.

Another side effect of trying to solder contaminated or incompatible materials is the tendency to apply more heat and "force the solder to take". As the materials aren't wettable the molten solder won't flow where you want it to flow. This can do more harm than good, because it may be impossible to burn off any contaminants anyway, and the component or the printed circuit board may be overheated and damaged in the process. Semiconductors may be harmed by applying excessive heat for more than a few seconds, and extreme heat applied to printed circuit board tracks can cause irreparable damage, because the tracks will be lifted away from the substrate especially on a delicate or badly designed board. You can avoid trouble by ensuring the surfaces to be soldered are clean and wettable to begin with.

## Getting Ready

As already explained, cleanliness in soldering is a major factor in obtaining successful results. A soldering iron stand usually has a sponge that is dampened – it wants to be quite damp but not running wet, so squeeze out any excess.

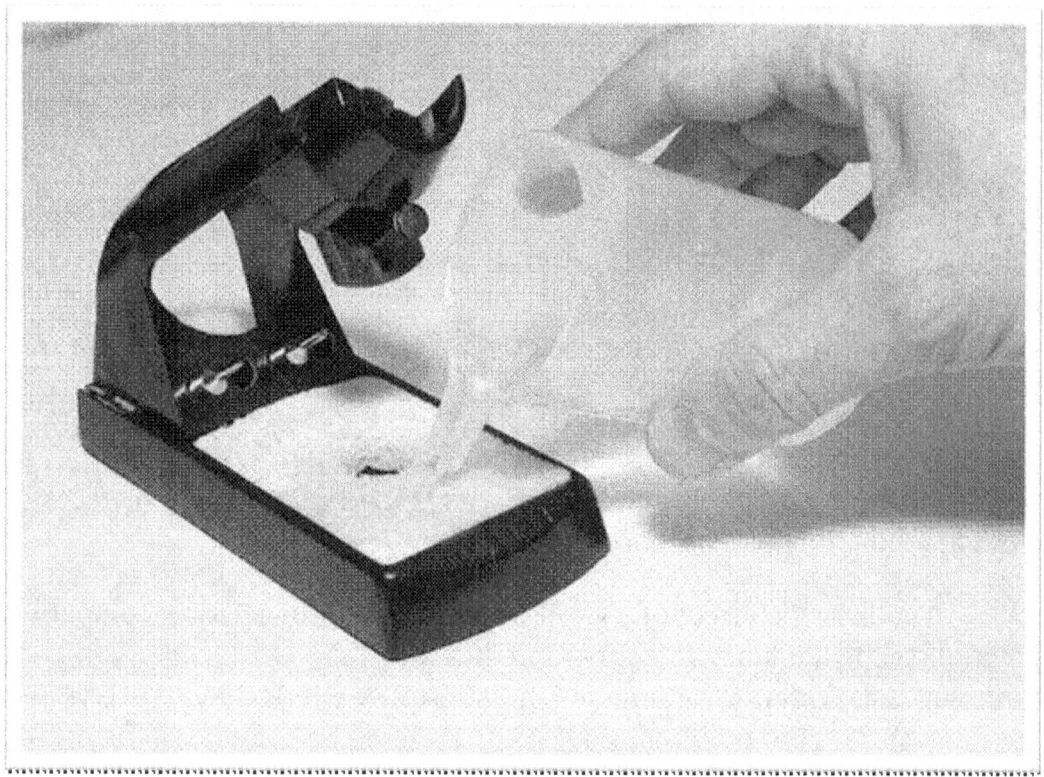

*^^ Dampen the sponge but don't soak it unduly. Note the hole, which gives an edge to wipe tip with.*

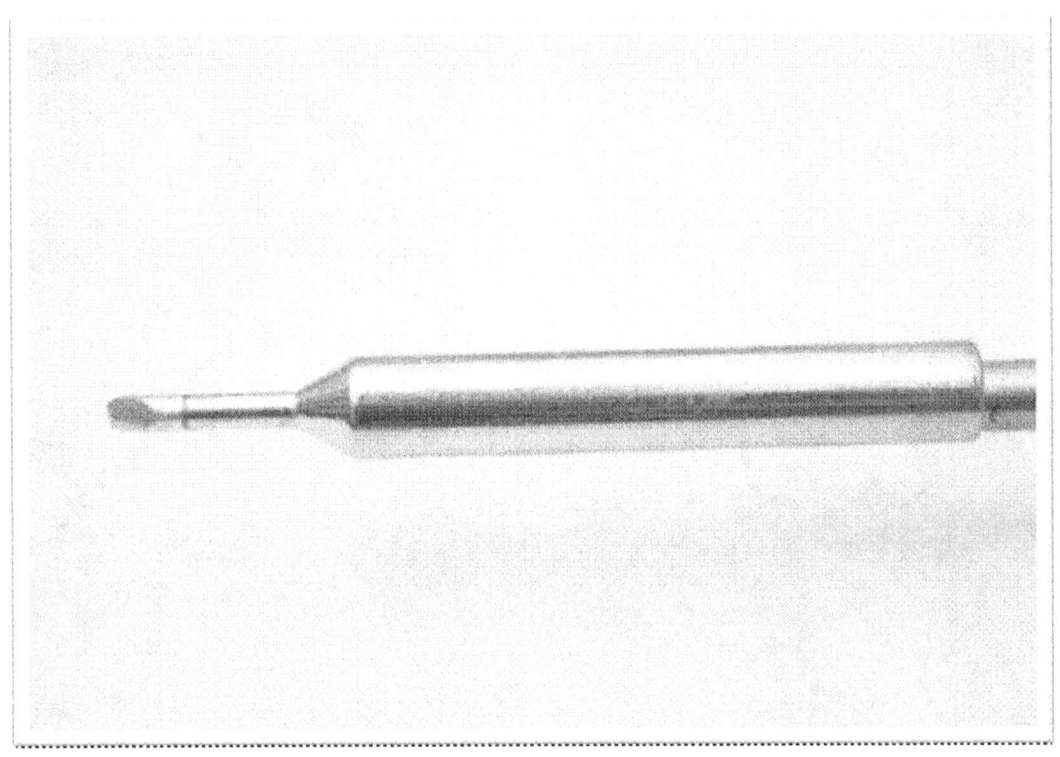

*^^ An Antex soldering iron tip. It simply slides onto the iron's heating element. Only the extreme end is wettable and will accept solder: the tip should be kept nice and shiny.*

Soldering iron bits are typically iron-plated to resist wear, then chrome-plated to prevent molten solder being deposited on them, with the extreme tip being "wettable" to work with molten solder. Before using the iron to make a joint, the hot tip must be "tinned" with a few millimetres of solder: *you should always flood a brand new tip with plenty of solder to tin it immediately, when using it for the first time.*

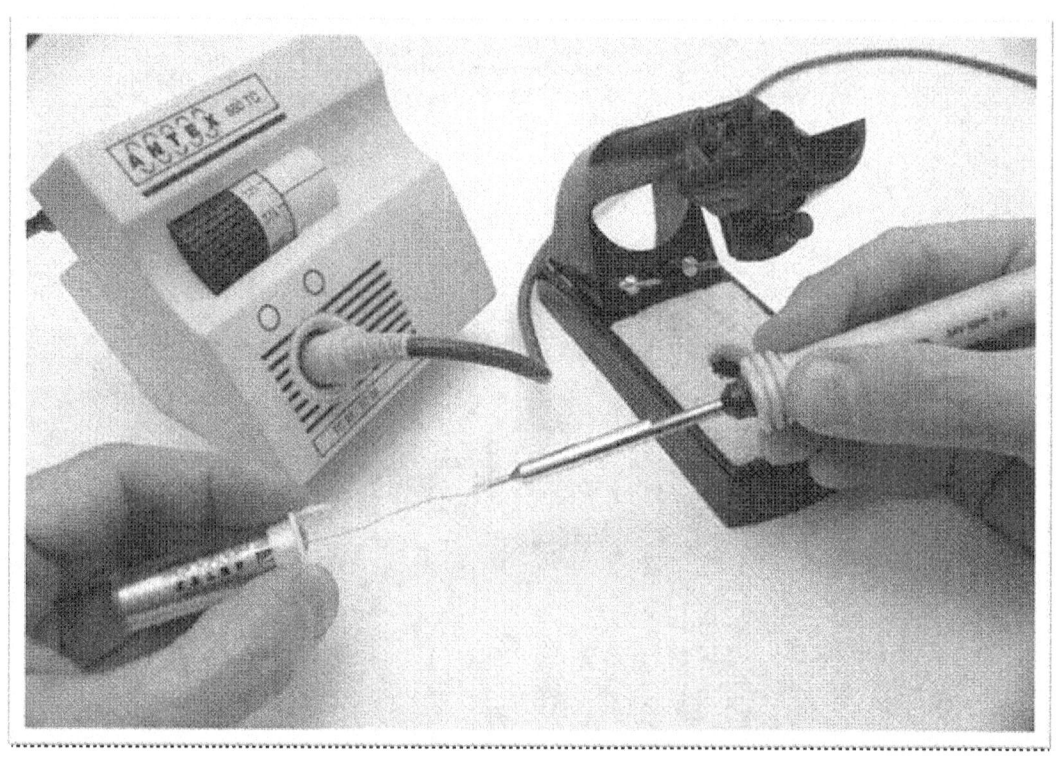

*^^ Tinning the tip ready for use: apply a few millimetres of solder to the hot tip.*

Wipe off excess solder using a damp sponge and it's ready to use. That's why sponges have a hole or well in them – the edge acts as a wiper and the hole catches excess solder.

A useful tip: after tinning the bit, just before using the iron it helps to re-apply a small amount of solder to a clean tip, to improve the thermal contact between the tip and the joint. The molten solder fills the void between the materials and the iron tip, to help transfer heat better so that the solder flows more quickly and easily.

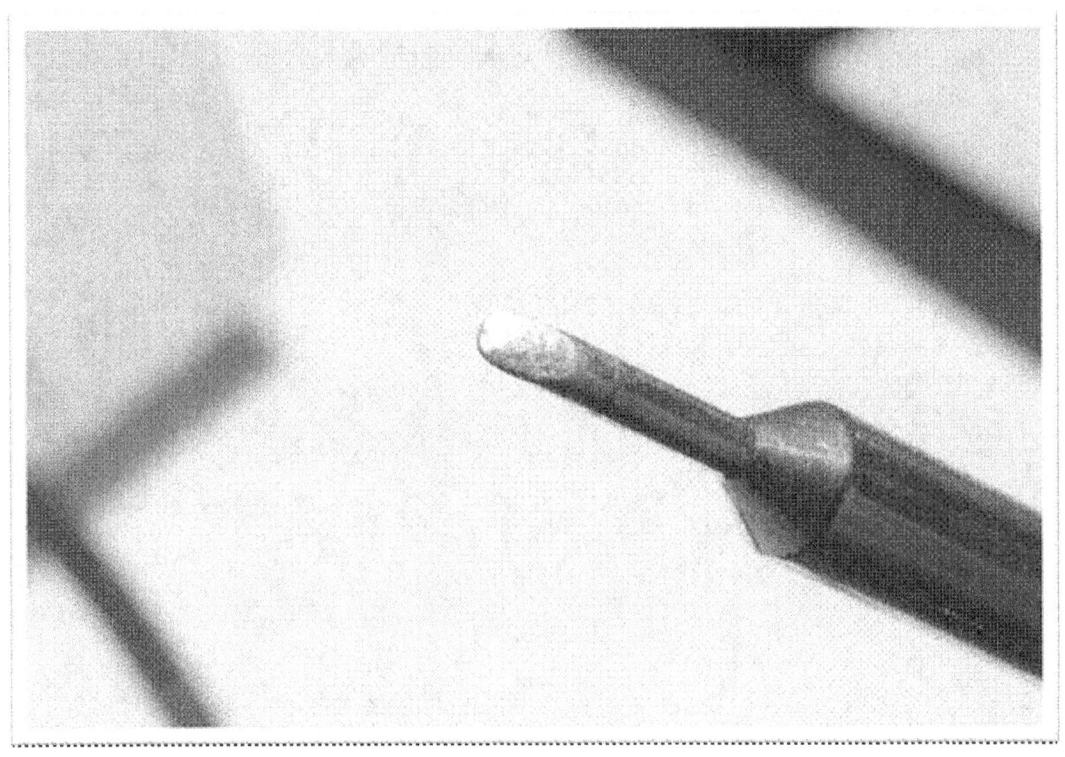

*^^ A tinned tip should look like this, ready to make the next joint.*

It's sometimes better to tin larger parts as well before making the joint itself, but this is not generally necessary with p.c.b. work. Small tinlets of special paste are available onto which you dab a hot iron - the product cleans and tins the iron ready for use. It's useful in extreme cases where the bit has some stubborn contamination.

I find special Tip Tinner &Cleaner products to be very useful: gently abrasive in action, they help to clean dirty bits and keep them in good condition. Use them for removing stubborn contaminants, but don't overdo it.

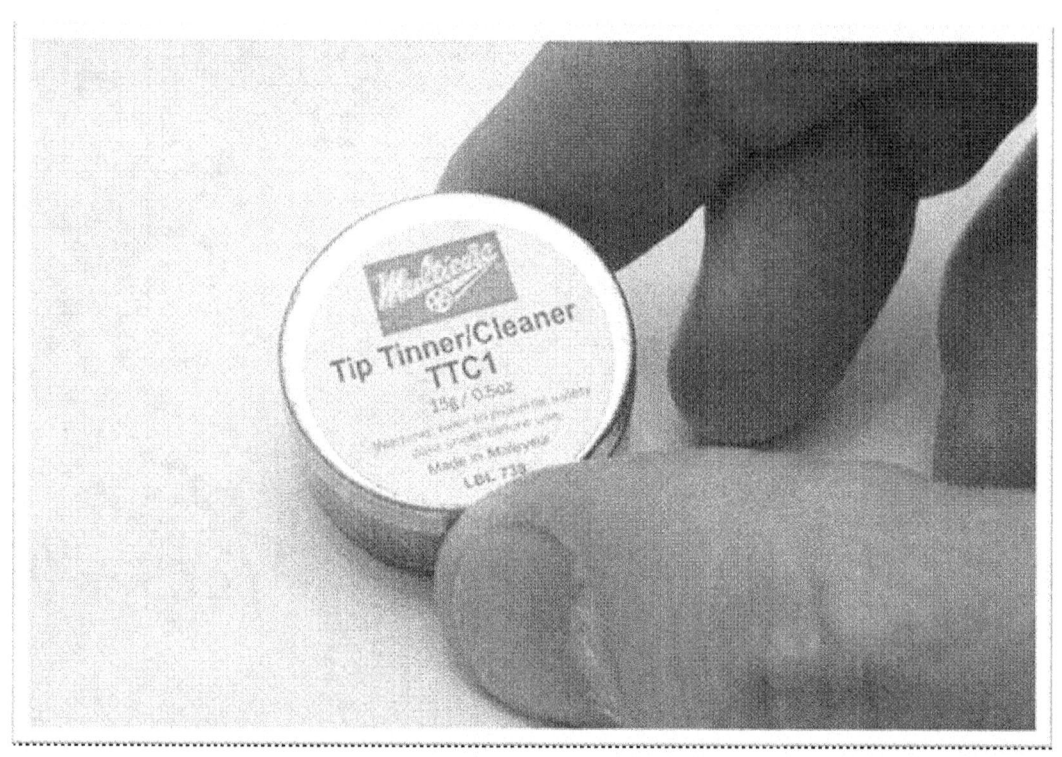

*^^ Tip Tinner and Cleaner is a useful aid to maintaining a soldering iron bit ...*

*^^ Just press the hot iron onto the solid paste and scrub it around a little. The tip will be cleaned, tinned, and made ready for use.*

The move to lead-free solders (see next chapter) has had some effect on the life of soldering iron bits, with increased wear and corrosion noted due to the higher temperatures and the fluxes found in tin-based solders. You can therefore expect bits to wear out over time. Once the iron-plating is damaged due to oxidation or erosion, the bit is fast approaching its end of life. Never use an abrasive or file to sand down a tip: the iron-coating will be damaged and the iron's core exposed, so the tip will soon be made useless due to erosion.

Having prepared the soldering iron tip ready for use, in the next chapter solder and fluxes are discussed.

## Solder and Fluxes

In recent years there's been a move towards using more environmentally-friendly materials in electronic products. EU legislation such as *Restriction of the Use of Certain Hazardous Substances in Electrical and Electronic Equipment* (RoHS) aims to reduce toxic heavy metals being sent to landfill. (Look for the RoHS symbol on equipment to indicate compliance.) Due to RoHS compliance, the electronics industry had to change the type of solder it uses in electronic production.

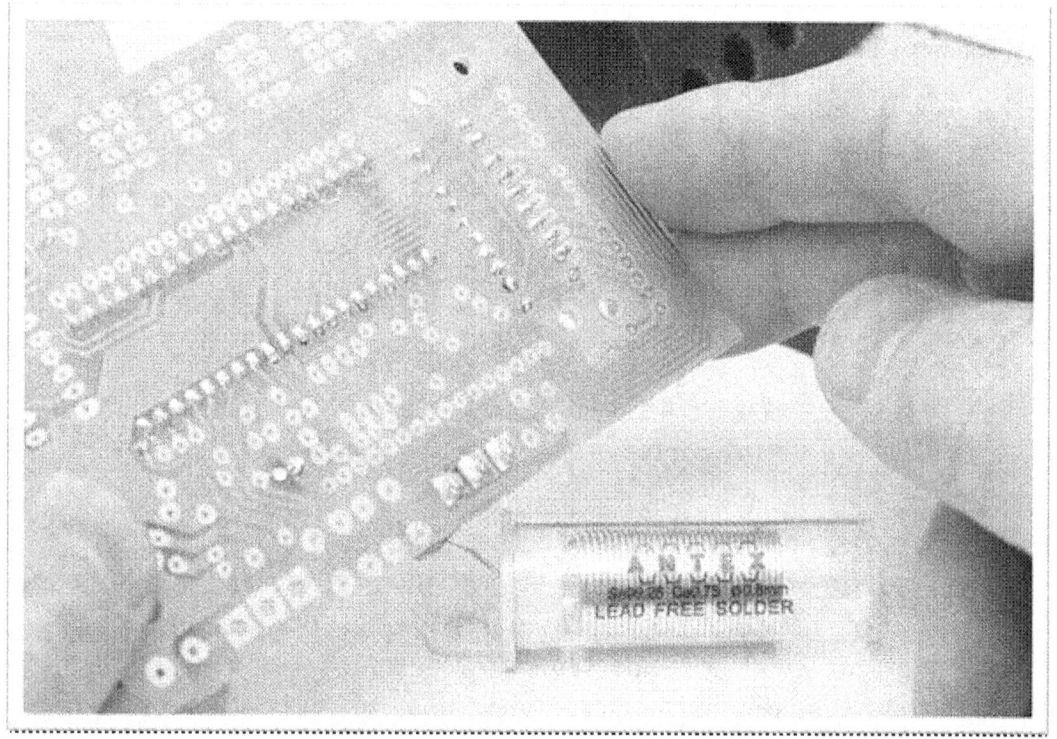

^^ *Electronics solder is supplied in reels or handy dispensers like this.*

Solder comes in various forms including solid bars or pellets for melting in small electric 'solder pots' used for treating the ends of wires with solder. Traditional general-purpose electronics-grade solder is in wire form – starting with so-called "60/40" which contains 60% tin (symbol Sn) and 40% lead (symbol Pb) and is sold in handy dispensers or reels. Although tin-lead solder is now banned in industry, there's nothing to stop the hobbyist from using it but best practice is to use lead-free solder in our work: my advice is to try both, and see which you prefer to work with. "40/60" tin-lead produces lower quality results but is slightly cheaper and perfectly acceptable in hobby circles.

Various diameters of solder wire are marketed. In the UK they're sold in Standard Width Gauge (SWG) sizes, typically as 18SWG (1.2mm) or 22SWG (0.7mm). The latter is fine for almost all hand-soldering of printed circuit boards or general electronics. For larger solder joints (e.g. larger switch or motor terminals), 18SWG solder would be better as more solder can be dispensed more quickly.

Lead-free solder is universally available and contains typically 99.7% pure tin and 0.3% copper (symbol Cu). It needs a higher melting point which makes it slightly more difficult to work with, but standard soldering irons will cope with it well. Antex lead-free solder (Sn 99.25 / Cu 0.75) is a good compromise at 0.8mm diameter and is sold in small dispensers.

Other solders are produced for specialist work, including aluminium solder (Alu-Sol®) and another solder variant used by professionals is Multicore "Smart" wire which contains a small amount of pure silver (symbol Ag). It produces very clean results and is often associated with SMD (surface mount devices), though some engineers also use it for routine p.c.b. work for producing the best possible finish by hand. As "Smart" wire contains lead it is not RoHS compliant.

An interesting variant is **Eutectic** solder, which is 63/37 Tin/Lead. It goes instantly from solid to liquid when melted and is particularly good for hand-soldering. An almost-equivalent lead-free product would be Stannol Flowtin TC or TSC solder.

## The low-down on fluxes

When melting solder with a soldering iron, oxides of metal are produced as a result of the high temperatures involved. Unfortunately, these oxides contaminate the metal surfaces being soldered, which interferes with the flow of molten metal and the production of a good quality solder joint.

All electronics-grade solder wire therefore contains an additive called "flux" which helps the molten solder to flow more easily over the joint. It does this by scrubbing away the oxides which arise naturally during heating, and it will often be seen as a pungent brown fluid bubbling away on the joint, accompanied by some fumes.

**Those coming into electronics from other industries should note that flux is already contained within "cored" electronics solder** and on no account should any acidic flux be applied separately before using the soldering iron. Plumbers, for example, apply flux paste to copper pipes before soldering them, but electronics-grade solder wire already contains a flux and extra flux is almost never needed. Electronics is no place for acid fluxes!

*^^ A close-up of electronics-grade multi-cored solder. Five cores of rosin flux can be seen running through it.*

For almost all electronic hand assembly, solder wire containing "Rosin flux" is used. Cores of flux run through the solder wire like letters running through seaside candy, and they prevent the hot area from being contaminated by oxides, otherwise solder would never flow properly and the result would be an incomplete and unreliable joint.

## Flux dispensers and Colophony

Flux dispenser pens are sold that allows special liquid flux to be applied separately onto a work area. These might be handy for difficult or challenging jobs to help solder to flow better: adding more flux this way won't do any harm and may help a solder joint to be made more quickly and reliably. In my hobby electronics, there's hardly ever been a time when I felt the need to apply extra flux but it's useful for some very tricky or demanding jobs.

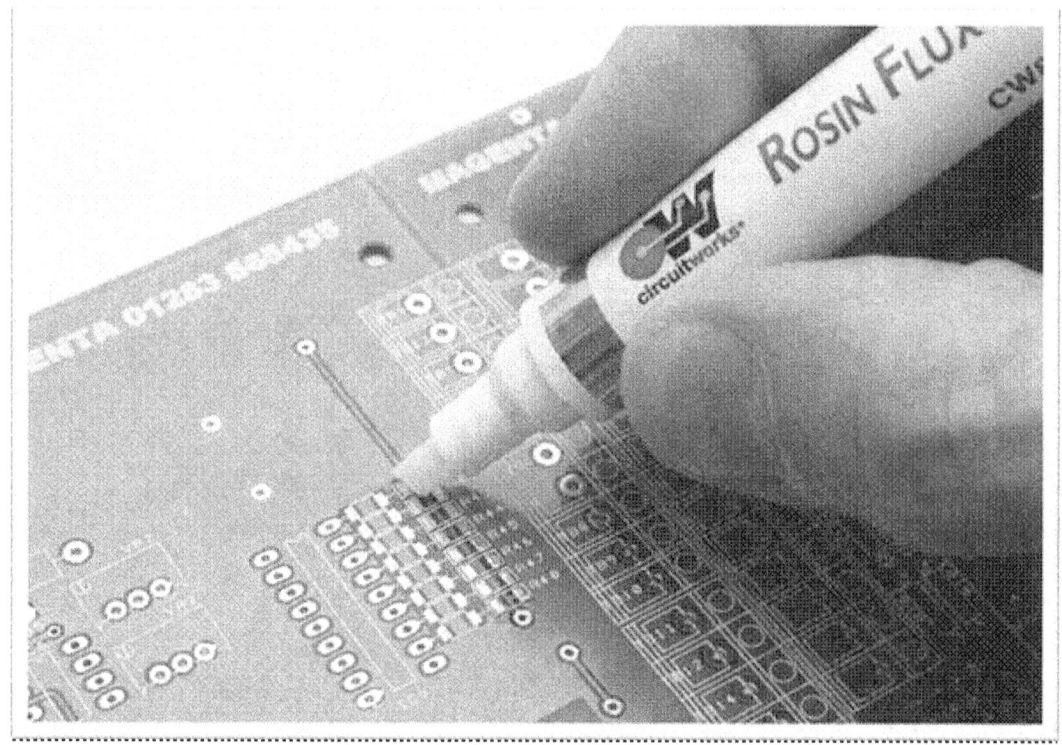

^^ *For more demanding work, a flux dispenser pen allows additional flux to be applied*

For example I've used specialist Chemtronics flux dispenser pens on tricky, extra-large solder joints involving very thick wires for lead-acid battery connections where I really struggled to make the solder flow properly. You might also use them in micro-electronic surface-mount work. The extra flux can only help, but for the rest of the time rosin-flux core solder wire contains sufficient flux and that's all that you'll need.

^^ *Colophony or rosin flux is available in small tinlets if you want to prepare your own liquid flux.*

One thing that many seasoned electronics enthusiasts will recognise is the distinctive smell of rosin flux: its intense woody pine smell is not unpleasant, and flux fumes themselves are not known to be harmful but solder smoke can be an irritant, especially if you suffer from asthma or other respiratory conditions. (I dealt with solder fume extraction earlier when soldering irons were outlined.)

Rosin flux is also known as **Colophony.** It's an amber resin that's glassy and brittle like sugar candy, distilled from the resins of conifers (mainly pine trees) and it's worth knowing about. For electronics use, Colophony is available in individual tinlets of solid resin (e.g. *Kolophonium* in 20g tins by Donau Elektronik GmbH, sold by Westfalia or Conrad).

As an aside, the resins of various trees are used in incense burners: placing a few fragments on a charcoal burner produces intense delicious fragrances of pine, frankincense or forest but you still wouldn't want to inhale the smoke or fumes directly.

Solids of Colophony can be dissolved to make a semi-liquid flux, for dipping or applying manually prior to soldering. To make your own rosin flux from solids of Colophony, chip off some fragments and crush them into a small tinfoil dish, then apply some isoproponal alcohol (start with 2 parts solvent to 1 part Colophony) and let it dissolve over about 20-30 minutes or more.

*^^ Chip off some solids of colophony and put them in a small tin foil dish*

*^^ Add e.g. x 2 volume isopropanol alcohol and allow the solids to dissolve. Experiment as needed.*

*^^ The resulting flux can be very sticky so handle with care and do not spill.*

In this form it can be applied directly with a brush or dropper to areas prior to soldering. In an open dish, after a day or two the solvent will mostly have evaporated to leave an extremely sticky resin that should be handled carefully: it's no co-incidence that Colophony is also used in adhesives and varnishes, and different grades of Colophony resins are used to treat violin or double-bass bows to add friction to bow hairs! Try experimenting with different ratios of solvent and making some up and storing it in an old nail varnish bottle with brush, or generally do what works for you.

Unused Colophony crystals can be reformed into their storage tinlet by warming carefully with a hot air gun, but don't overdo it. You can also use flux cleaners or PCB solvent cleaners if you want to remove traces of excess rosin flux after soldering: these are left behind as a brown deposit and are otherwise harmless.

The subject of using your soldering iron to raise the temperature of the materials is discussed next.

## Temperature flow

This aspect takes a bit more understanding, but with practice you'll soon understand how temperature flows within a workpiece being soldered and how you have to harness it properly. After ensuring all parts are clean and the soldering iron is ready to go, the next step to successful soldering requires that the **temperature** of **all** the parts is raised to roughly the same level, before solder can be applied.

Imagine the most basic task of soldering a simple resistor onto a printed circuit board: the copper p.c.b. and the resistor lead should both be heated *together* so that the solder will flow readily over the joint. Later I'll show the precise stages step by step.

A beginner will often mistakenly just heat one part of the joint (e.g. a resistor's wire protruding through a printed circuit board) and hope that the resultant "blob" of solder will be enough to tack everything together. That's completely wrong, because the remaining metal in the joint is cold when molten solder is flooded on to it. The joint will be weak, incomplete or unreliable. Flux will not have flowed properly either, so the joint could be contaminated internally.

Another beginner's mistake is to use a soldering iron to carry blobs of molten solder over to the joint, as if to daub solder over it. The secret of success is to control the iron accurately and apply the hot tip onto the workpiece so that it's in contact with *all the parts*. Within a fraction of a second, heat will conduct from the iron and raise the temperature of the entire joint, after which solder can be melted over it. Remove the iron and let the joint cool naturally. It takes some practice with your chosen soldering iron tip to obtain the best heating action, making sure the tip is clean and tinned properly to begin with.

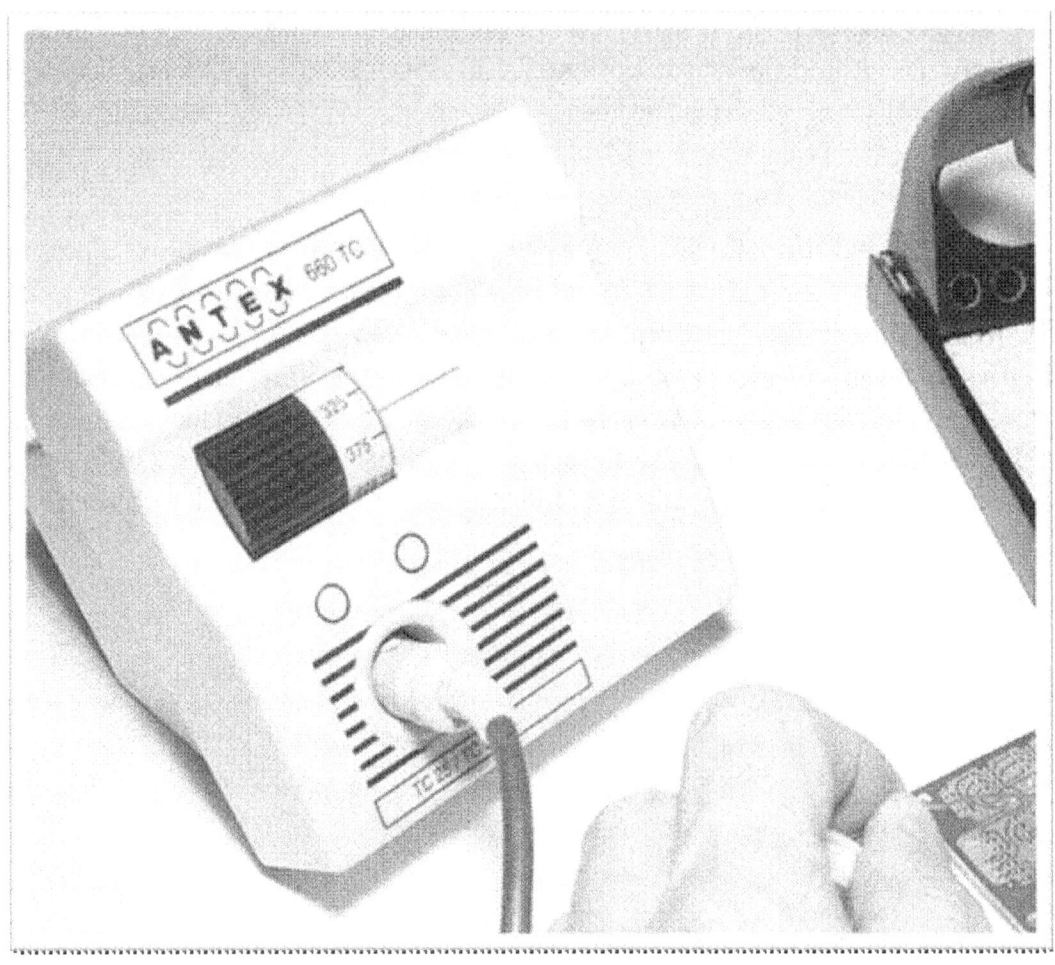

*^^ Set a typical temperature of 330-350 °C (626-662 °F) or more, if you have a soldering station.*

The melting point of traditional tin-lead solders is about 190°C (374°F) but lead-free tin-based solders require higher temperatures, having melting points of typically 201-227 °C. As Antex reminds us, the melting point is *not* the temperature of the soldering iron tip: instead you should set a temperature that ensures the solder melts *instantly* onto the tip. Fixed-temperature soldering irons have no adjustability but they'll cope just fine with either type of solder. A soldering station usually has a variable control that gives more control in different circumstances. In practice the iron tip temperature should be set for typically 330-350 ℃ (626-662 ℉) or maybe a little more if using lead-free solder.

## Now is the time...

The next diagram shows what would happen if you applied a hot soldering iron to an imaginary metal block. In step (1) heat travels out of the iron into the cold metal block, which starts to warm up starting at the edges. Gradually (2), the whole metal block's temperature rises until the middle of the block finally rises in temperature as well. In effect heat is now travelling back "towards" the iron tip, until finally (3) the whole block is at the same temperature as the iron. At this point solder could now be applied.

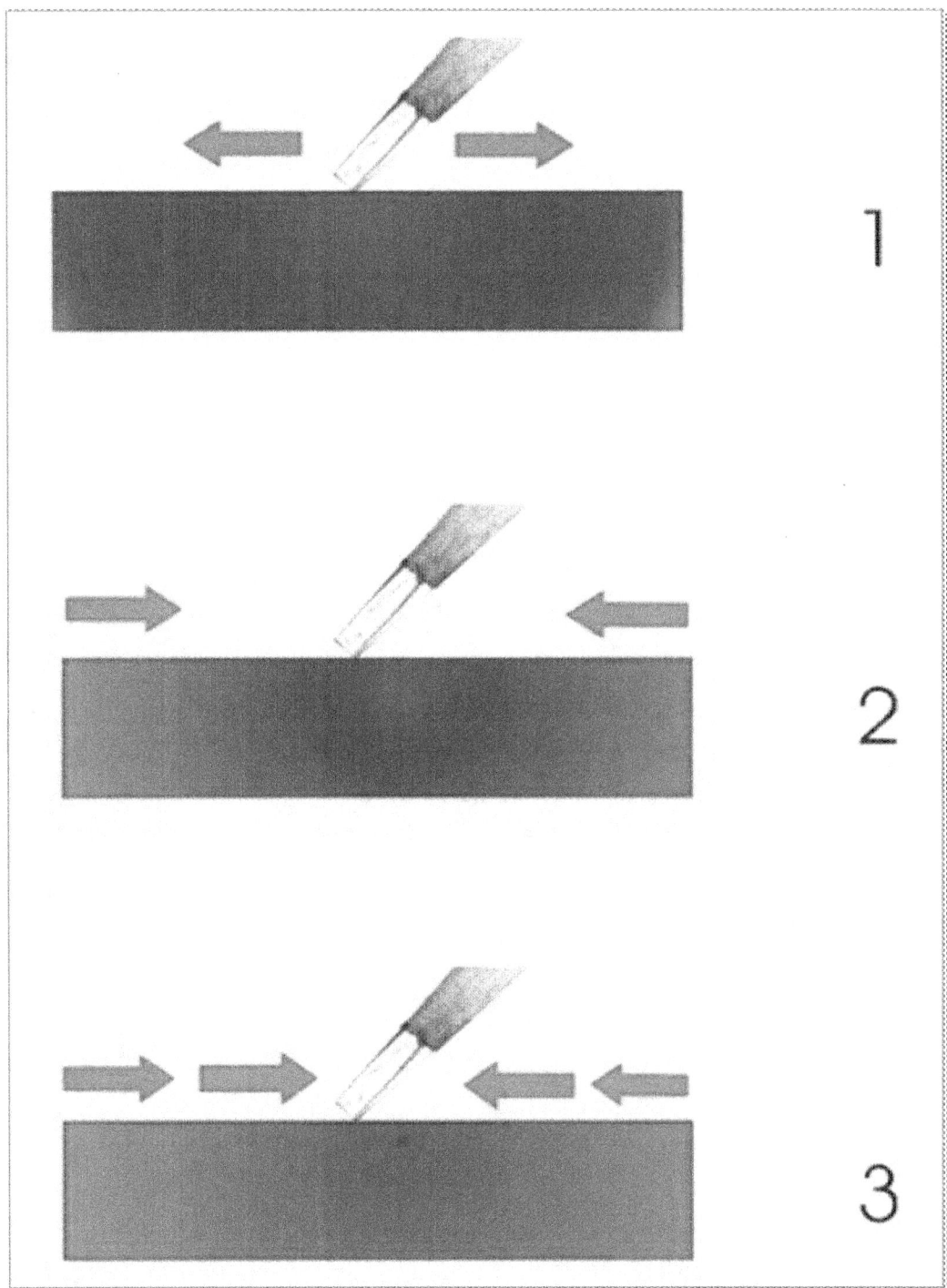

*^^ How metal in the joint actually "sinks" heat away from the tip to begin with. Then heat moves back towards the tip until the solder's melting point is finally reached.*

You'll notice this effect as a time delay when soldering any joint. The "working area" of the iron's tip gets **cooled** to begin with, because the metal in the rest of the joint is sinking heat away from where it's most wanted. Only after the whole joint has risen in temperature can solder be melted onto it.

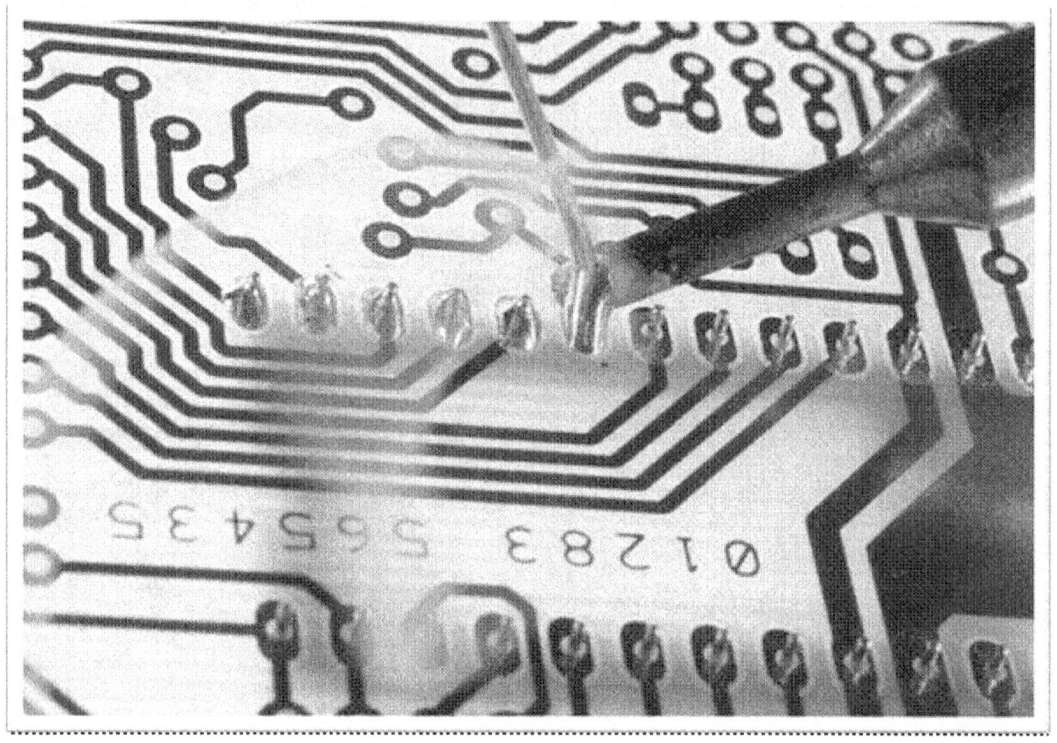

*^^ Soldering a small p.c.b. joint*

With experience, you'll get a feel for how long it takes before you can apply solder. The more metal that is present in the joint, the longer period that heat must be applied for. A small p.c.b. joint takes well under one second to complete. A large metal terminal could take quite a few seconds or more to heat up. As I explained earlier, higher power (wattage) irons cope better with larger workpieces because they recover more quickly and are more "unstoppable", making it easier for them to heat larger workpieces without cooling down so much.

If you apply solder too early, it won't melt properly and the result will be a grey, crystalline joint caused by the solder's melting point temperature not being reached and the flux not having flowed properly. Semiconductors must be soldered as rapidly as possible as they are heat-sensitive, but they're a lot more robust than they used to be.

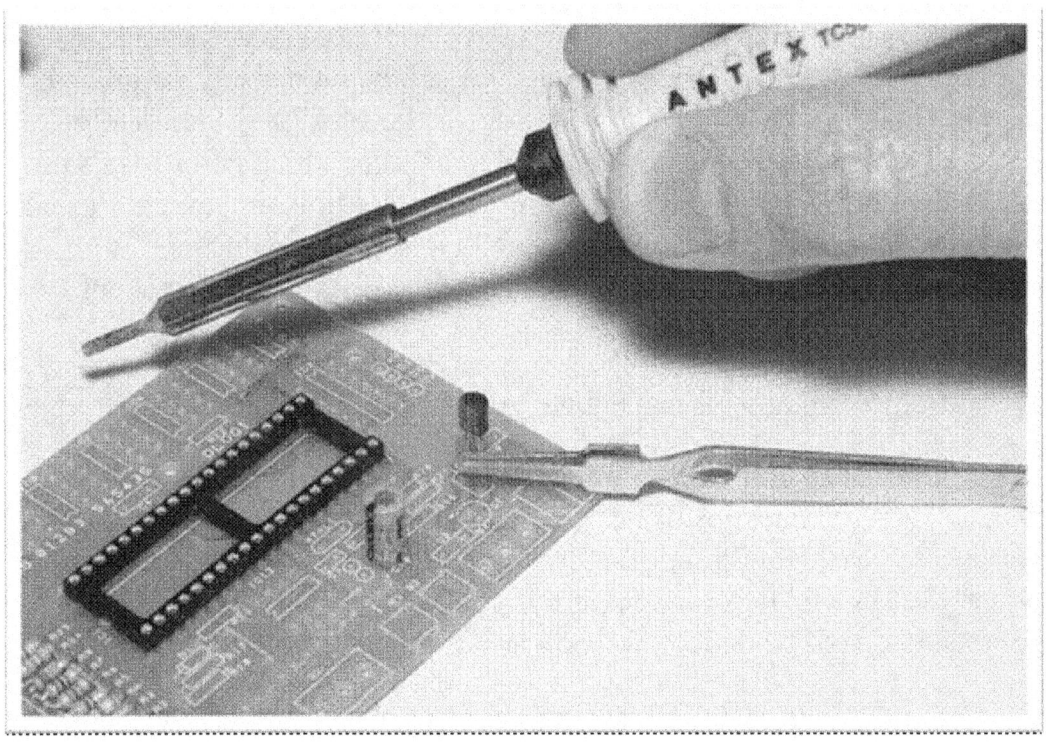

*^^ A clip-on Antex heat shunt fitted to a transistor leg, helps prevent thermal damage due to overheating when soldering it in place. Less essential these days, but beginners find them re-assuring.*

Until they have gained some practice, novices sometimes buy a small clip-on heat-shunt (photo) which resembles a pair of aluminium tweezers. In the example of, say, a transistor, the shunt is attached to one of the leads near to the transistor's body. Any excess heat then diverts up the heat shunt instead of into the transistor junction, thereby avoiding the risk of thermal damage. Applying far too much heat may destroy the part or damage the p.c.b. foil which may lift away from the board.

In due course constructors learns to judge how much solder should be applied to any particular joint. An ideal p.c.b. joint is very slightly concave in shape. If not enough solder is used, the result may be an incomplete joint which may cause an intermittent fault later on. An excess of solder – shaped like a ball bearing - is an unnecessary waste and in extreme cases may cause short circuits, especially on densely-populated boards. There is no need to add more solder "for luck". Professionally-produced p.c.b.s have a green *solder resist* coating which helps to ensure that solder does not stray onto adjacent pads. As a finishing touch, I usually spray the solder-side of a circuitboard with aerosol spray lacquer afterwards. It keeps the solder joints nice and shiny and helps prevent corrosion.

Some components can create hazards during a soldering operation:

**Coin cells and button batteries** are commonly used as power, clock or memory backups. If heated excessively they can explode without warning due to the build-up of internal pressure. Spot-welders are used in industry to connect tags to them, but if you need to solder wires to such a cell then it should be done as quickly as possible.

Some **memory back-up capacitors** or **electrolytic capacitors** remain energised for a while even when the circuit is powered down. Molten solder is a perfect electrical conductor and in some cases the component's contacts could be shorted during the soldering (or desoldering) operation. If molten solder shorts it out then the arcing may cause globules of solder to be spattered outwards without warning, potentially risking eyesight damage.

Always ensure that powered components are electrically inert and discharged before soldering. Cells, batteries and battery packs should not be accidentally shorted during the soldering process, to avoid arcing and solder spatter. Note that electrolytic capacitors can also explode after a while if reverse-connected, so observe polarity closely.

Let's now consider the practical stages of soldering components and wires successfully.

## Soldering Step by Step

Earlier I explained the individual factors that affect the quality of a solder joint. These are:

- **Cleanliness** – dirt or impurities drastically hinder good solder coverage.
- **Temperature** – the right level to enable the solder to flow freely!
- **Time** – apply heat for just the right amount of time!
- **Adequate solder coverage** – enough to form a good joint without touching neighbouring areas.

These rules apply whether soldering a p.c.b. or performing other tasks such as interwiring (hooking everything together with connecting wire).

We'll now summarise the stages of making a typical solder joint – soldering components onto a printed circuit board (through-hole soldering). Most people insert components into the circuit board and simply splay the wires out to hold them in place under spring tension. I find it best to snip excess wire leads off at this stage, to improve accessibility.

- All parts must be bright, clean and free from dirt and grease.
- Try to secure the work firmly to stop parts moving around.
- "Tin" the soldering iron tip with a small amount of solder. Do this immediately, with new tips being used for the first time.
- Wipe the tip of the hot soldering iron on a damp cellulose sponge to remove excess solder or contamination.
- Many people then add a tiny amount of fresh solder to the cleansed tip just before using it.

- Heat all parts of the joint with the iron typically for under a second or so, until it's heated throughout.
- While heating, then apply sufficient rosin-core solder to form an adequately-covered joint.
- It only takes a second or two at most, to solder the average p.c.b. joint this way.
- Do not move parts until the solder has cooled.
- Remove and return the iron safely to its stand.

This special photo sequence illustrates these stages. It's best to start with the smallest, fiddliest parts first when soldering a blank p.c.b., because that's when you've got the most access on the board. Accessibility will be reduced as more components are added, so we'll start with a simple wire link on a professionally designed p.c.b.

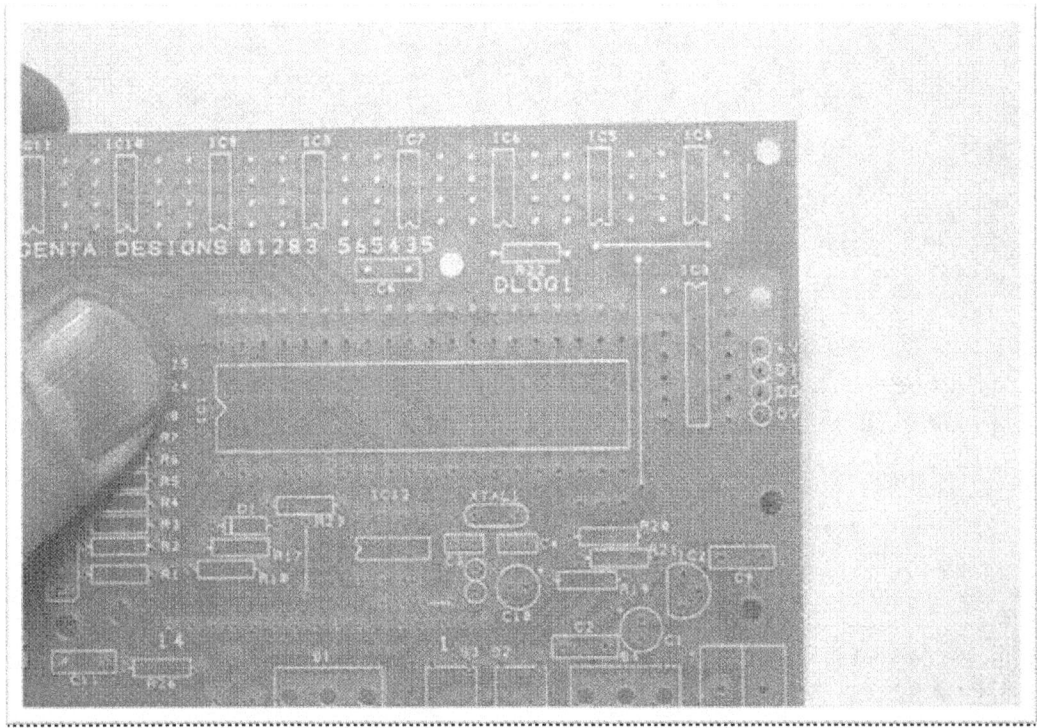

^^ *A typical professional blank p.c.b. – silk-screen printing shows what goes where. The underside has been treated with a green solder resist coating, and the solder pads are ready-tinned to help with soldering.*

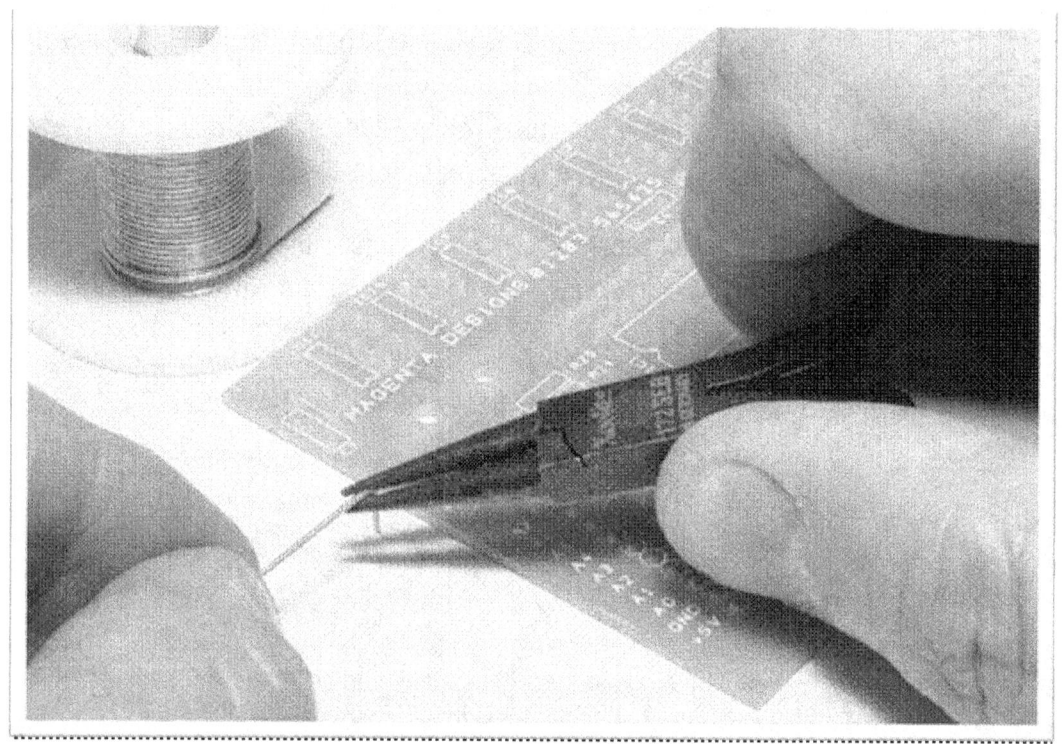

^^ *Preparing a wire link for soldering – cut off some tinned copper wire and bend the ends to fit the p.c.b. correctly. Round-nose pliers (shown) are perfect for this, but ordinary electronics or "radio" pliers will do.*

^^ *Wires are prodded through the holes in the board, then turn it upside down to view the solder side. You can then "spring" or splay the ends apart slightly, so they are held in place while you solder them.*

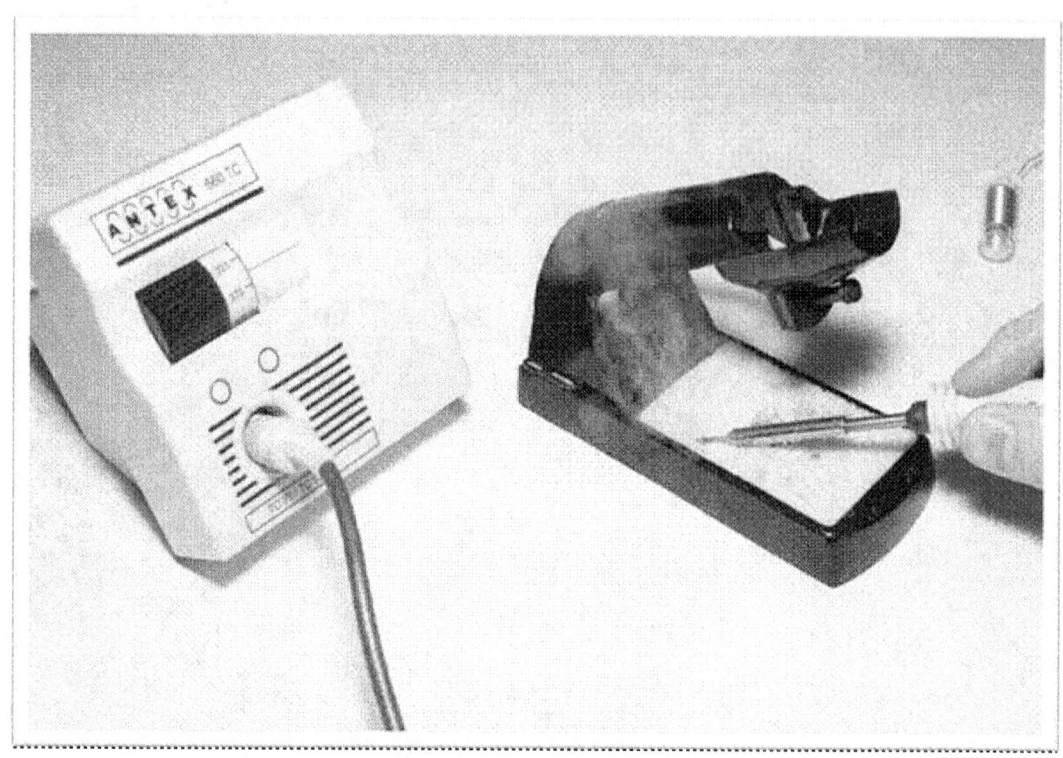

^^ *Wipe the hot soldering iron on the damp sponge to clean the tip. Do this periodically when contamination, flux deposits etc. build up on the iron to keep the tip nice and shiny. Tip Tinner & Cleaner helps too.*

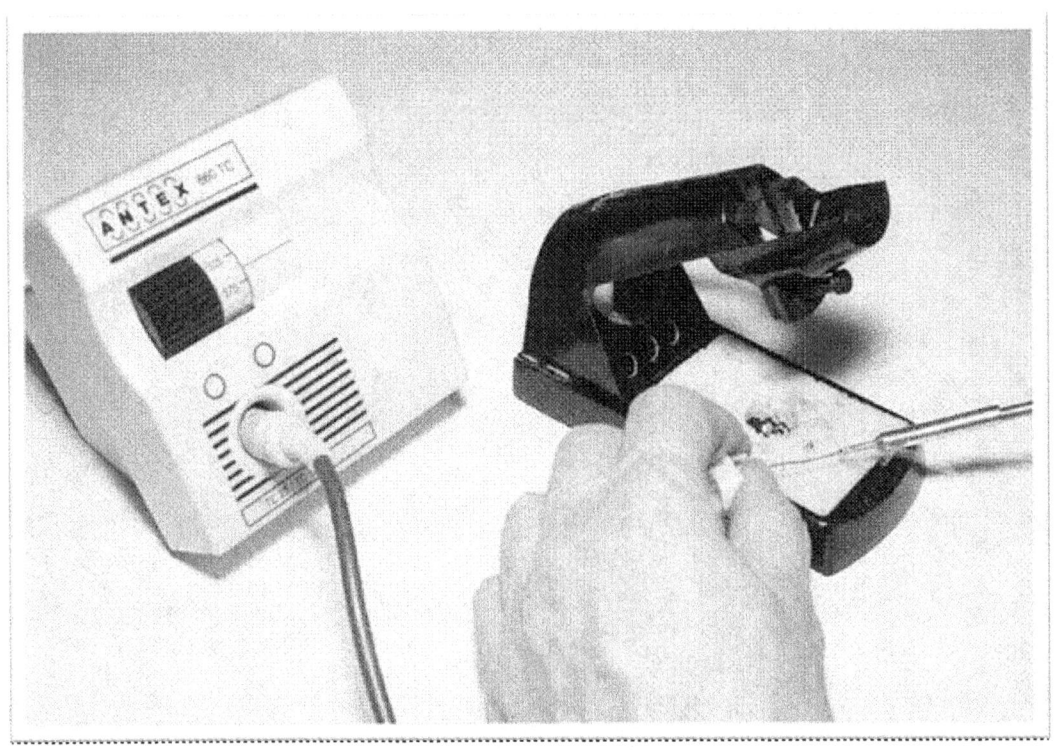

*^^ It often helps to dab a tiny amount of solder wire onto the tip, to improve heat transfer.*

^^ *Then apply the soldering iron to heat both the solder "pad" and the wire end at the same time (say <1 second). Apply a few millimetres only, of solder. Then remove the soldering iron immediately and allow the joint to cool down by itself. The green solder-resist coating ensures solder doesn't flow onto neighbouring pads.*

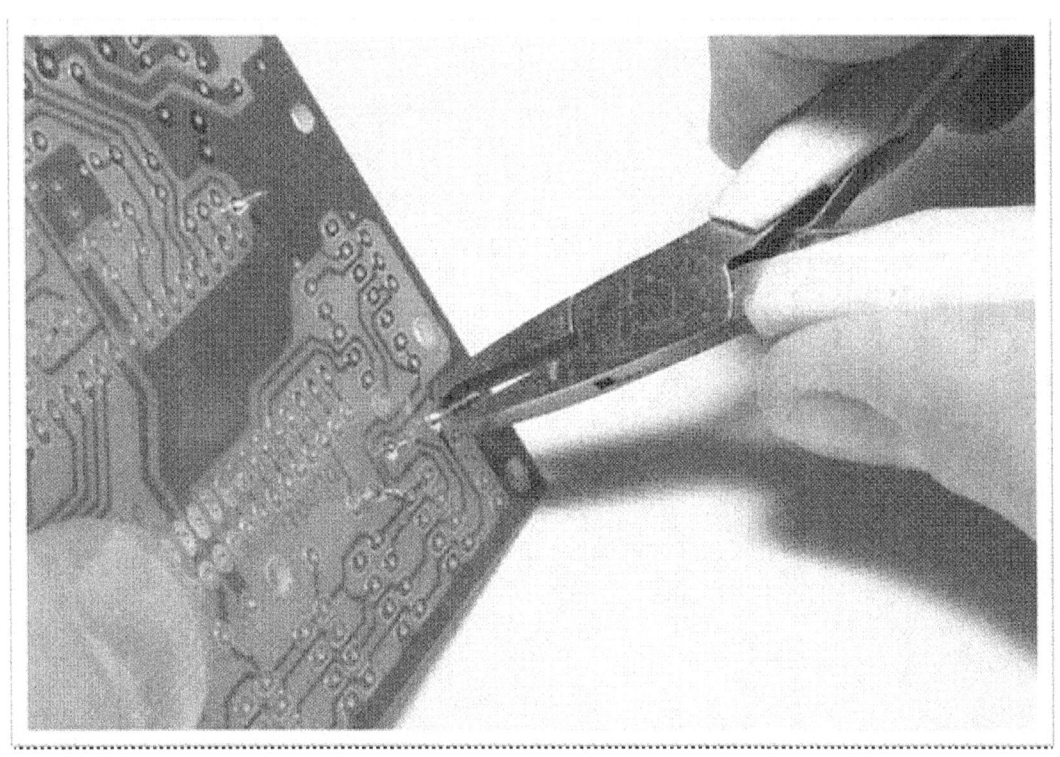

*^^ These "end cutters" can then be used to snip off excess wire after soldering....*

*^^ ... but  ordinary electronics "side cutters" are fine for trimming or snipping wires.*

*^^ The ideal solder joint should be smooth and slightly concave, quite shiny, not dull or crystalline-looking nor ball-bearing shaped.*

Try to get consistent results in your soldering, but don't worry too much about the lack of uniformity – this is soldering by hand after all, not by precision machinery, and some slight variation is OK provided the solder coverage is good and the soldering is generally clean and effective. You'll improve with practice, that's for sure!

*^^ A large i.c. socket can be added next...*

^^ *Go along the rows of pins and solder them one at a time. See how simple it is! Heat both the solder pad and pin with the iron, and dab on a little solder wire to ensure full coverage. Ensure the socket is flush against the board.*

*^^ As hardly any metal is involved each pin should take well under a second to solder at most.*

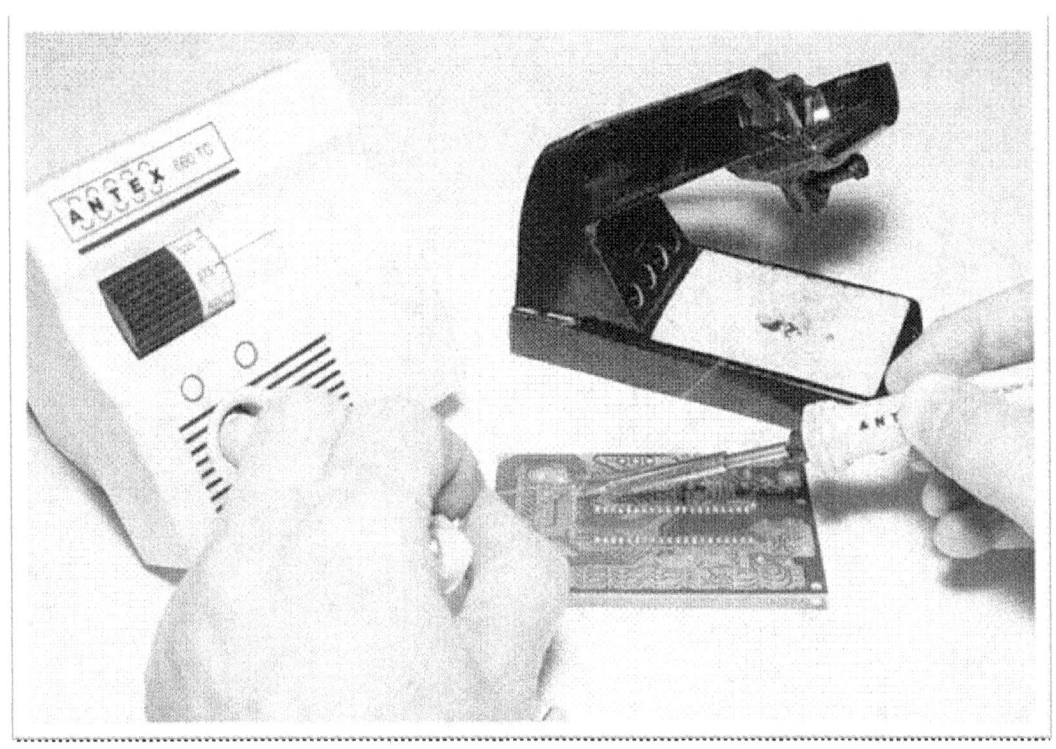

^^ *Continue until the i.c. socket is soldered.*

Once the smallest parts are soldered into place, you can continue to solder the remaining components. It's easiest to handle the smallest so-called "discrete" parts first while you still have plenty of room on the board. I usually solder resistors and capacitors next. The principle of soldering them is just the same as a simple wire link: insert them from above till they are flush on the board, then splay their wires a little to hold them in place, and preferably snip off at least some of the wire to give you more access.

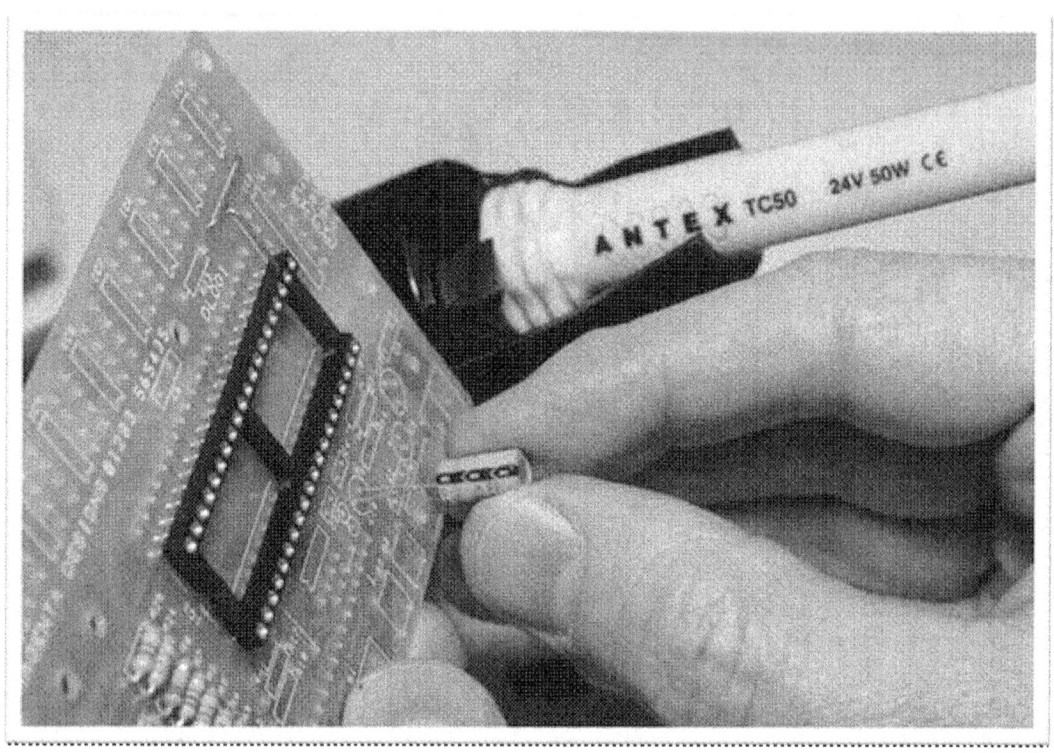

^^ *Continue by inserting "discrete" components from above, splay their wires out underneath to secure them in place, then snip and solder the joints exactly as before. Some parts like this blue electrolytic capacitor are polarity sensitive (note the – sign), and must be inserted the right way round. Same is true of every semiconductor.*

^^ *This row of ¼ watt resistors was next. I like to show the colour codes all the same way round, with gold or silver (tolerance bands) on the right for consistency.*

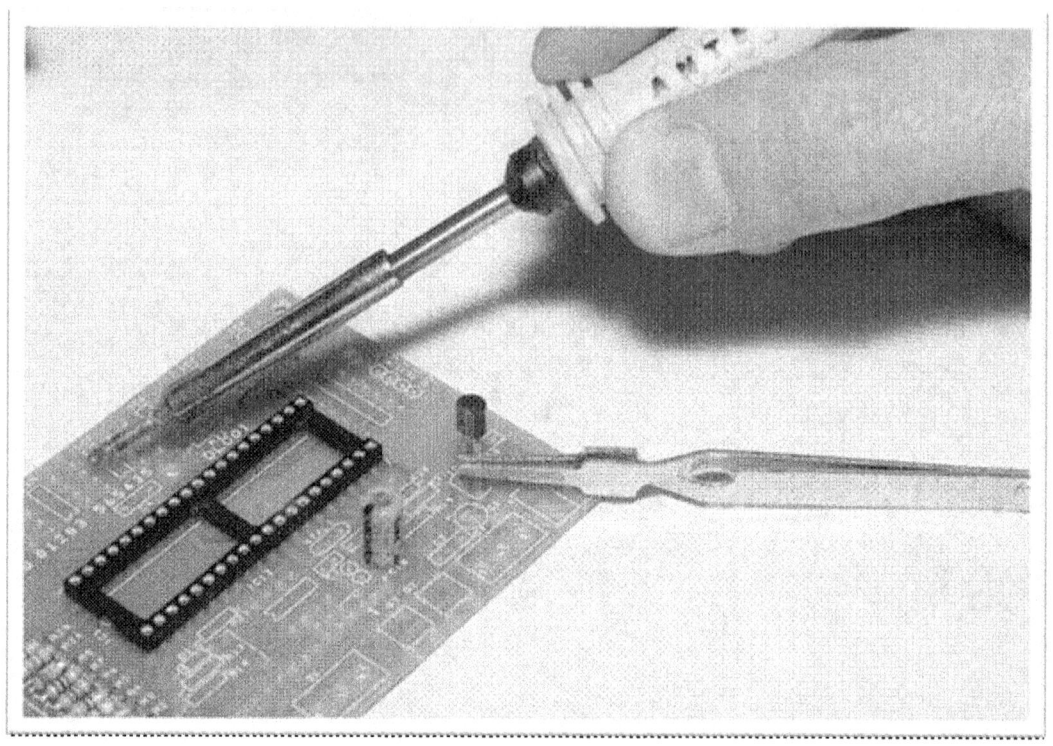

^^ *This transistor was next. Solder it quickly to avoid damage, and observe polarity correctly, so it is the right way round on the p.c.b. The heatshunt is optional.*

*^^ Unusually, this toggle switch fitted directly onto the board as well which saved wiring it up. Neat idea!*

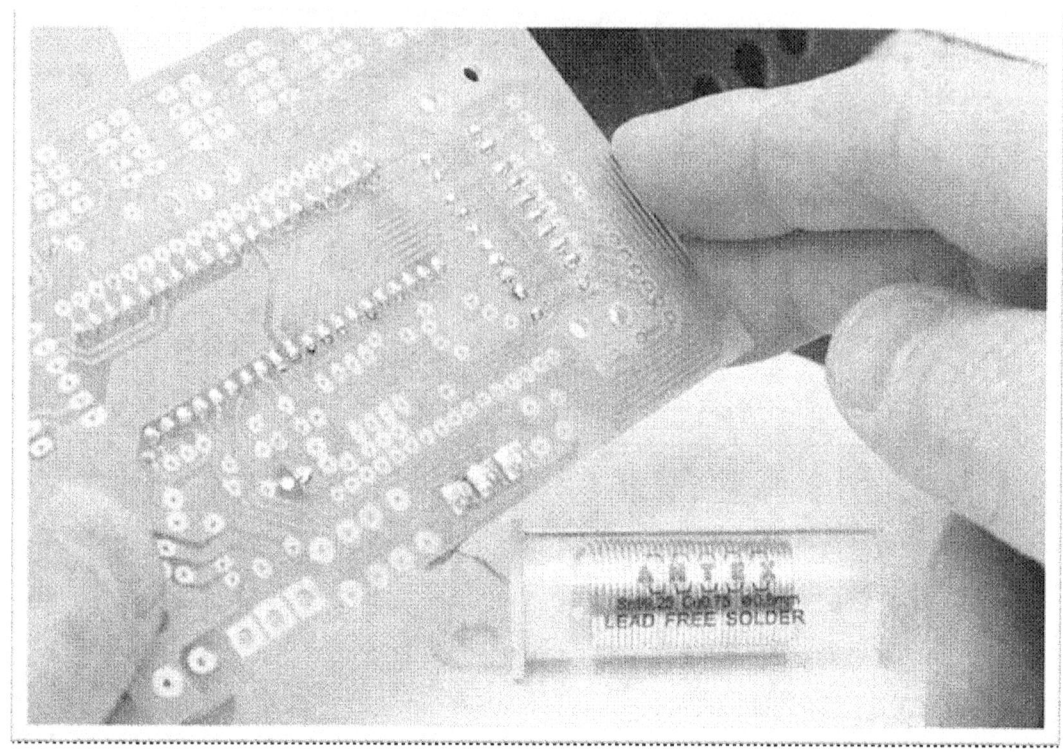

*^^ As there is comparatively more metal to heat up, it'll take longer to solder the switch terminals, and you'll need more solder as well. Thicker gauge solder is useful at such times. Allow say 2-3 seconds to solder each terminal.*

Don't forget to clean the soldering iron tip on its damp sponge every now and then, to ensure the bit is kept clean and shiny. Later on I'll show you how to correct any problems by desoldering using various techniques. Next, we'll move on from printed circuit board "through-hole" soldering and look at how to handle wires and leads.

## Interwiring – get hooked up!

With practice, through-hole soldering of p.c.b.s will become second nature. There's no substitute for tackling some soldering jobs though, particularly trying a simple kit based on a quality p.c.b. which will boost your confidence enormously.

Most electronic devices need connecting up to external components such as battery packs, speakers, l.e.d.s or switches. Usually, multi-stranded connecting wire is used to connect circuit boards and external parts together. Unlike single solid-core wire, *multi-stranded wire* is flexible and vibration-resistant. Hobbyists mainly use 7/0.2mm wire (7 strands, each 0.2mm diameter) for low-voltage hook-ups although much Chinese equipment uses much thinner wires than this. So let's look at some aspects of soldering this type of wire.

In a separate photo sequence I show how a potentiometer (a panel-mounted variable resistor) and a light-emitting diode (l.e.d.) are connected using multi-stranded wire. The same principles of soldering apply to most other components including panel-mounted switches, loudspeakers, buzzers, audio sockets and more.

Components usually have terminals or "tags" to which wires can be soldered. Start by ensuring the component's tags are clean: otherwise solder will not wet properly and the joint will be impossible to solder, so all contamination must be removed. This is especially true of parts that have been in storage a long time. The connections often oxidise or blacken, so clean the solder tags with e.g. an abrasive glass-fibre brush, or a needle file or abrasive paper. Using a glass fibre brush was shown earlier in "Cleanliness and Tinning the Bit".

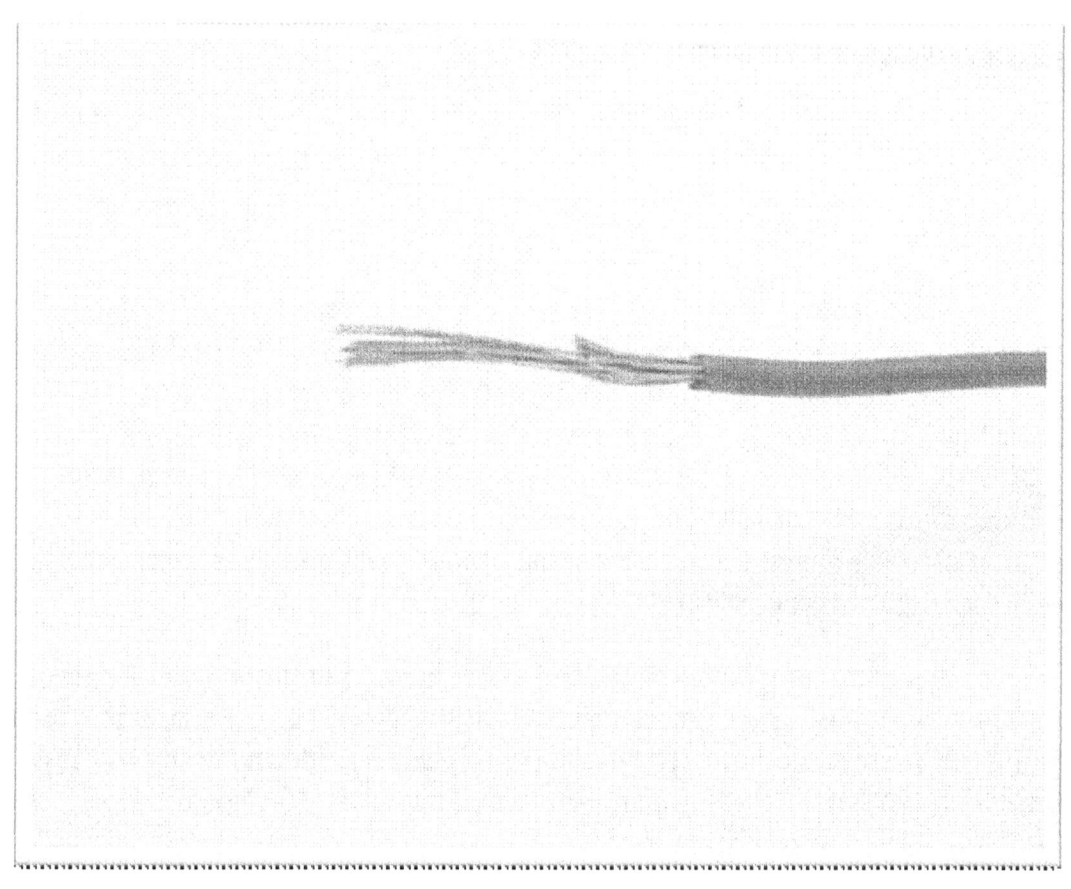

*⋀ How not to strip insulation off wire: some of the cores have also been cut – avoid doing this!*

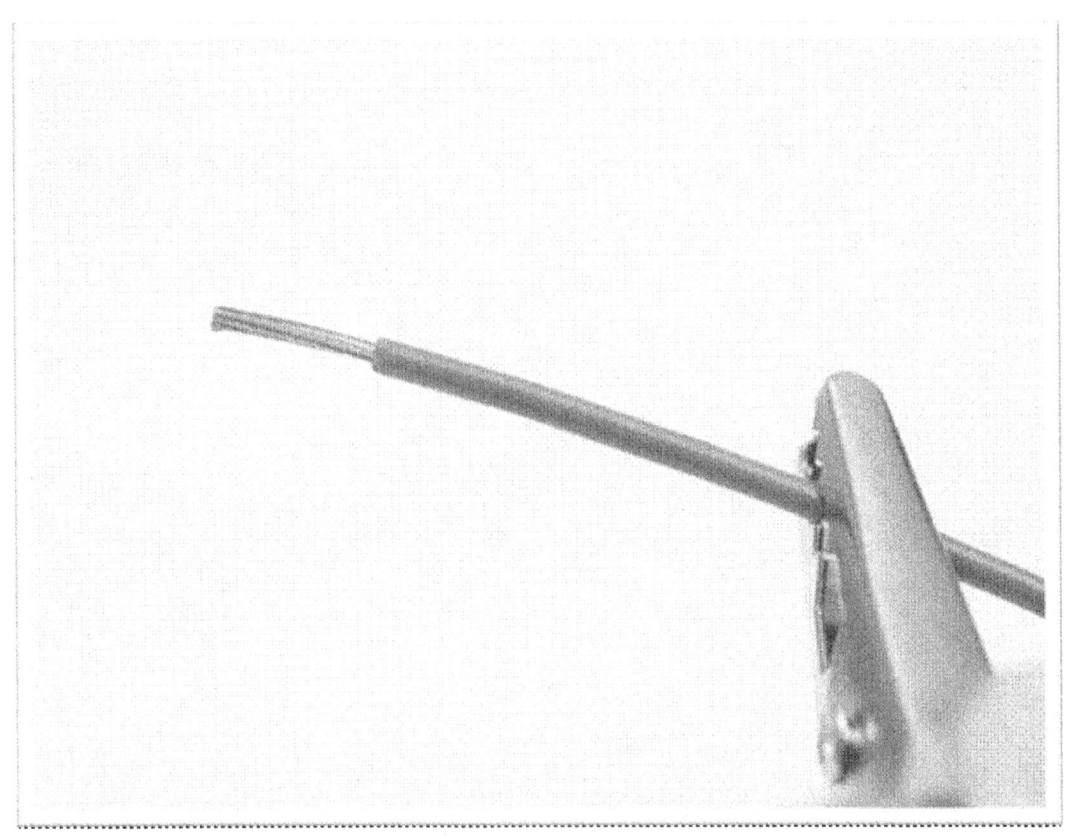

*^^ Gripping a wire end in a "Helping Hand" croc clip can help with soldering.*

After stripping a short length of insulation from the connecting wire, there are two ways to solder it to a component's solder tag. The first way is to *"tin" the stripped wire end to solidify it* - just heat it with the iron and melt a little solder on it, and let it cool. Poke the wire end through the solder tag, apply the hot iron to both parts and solder them together using a few millimetres of solder.

Although the assembly doesn't hold itself together so well during soldering (consider a Helping Hands jig if needed) this is quicker and easier to make and also easy to desolder again, and is perfectly adequate for most joints of this kind. The majority of commercial wire joints seem to be made this way.

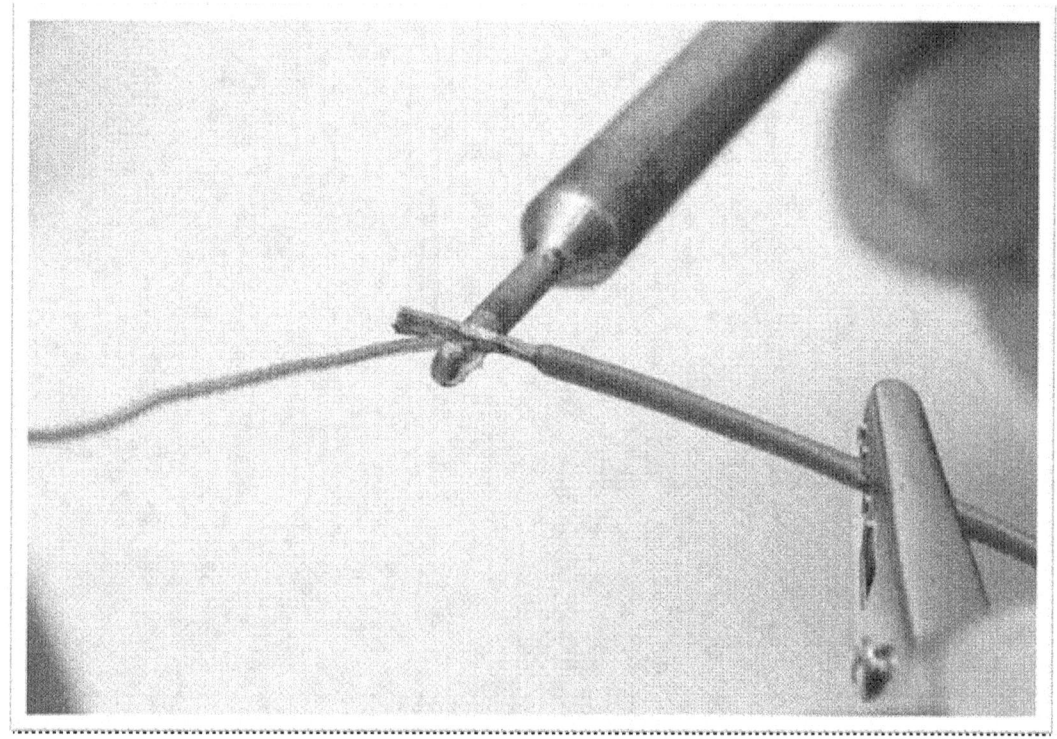

*^^ Apply a hot iron and solder, in order to "tin" the wire ends. This makes them into a solid.*

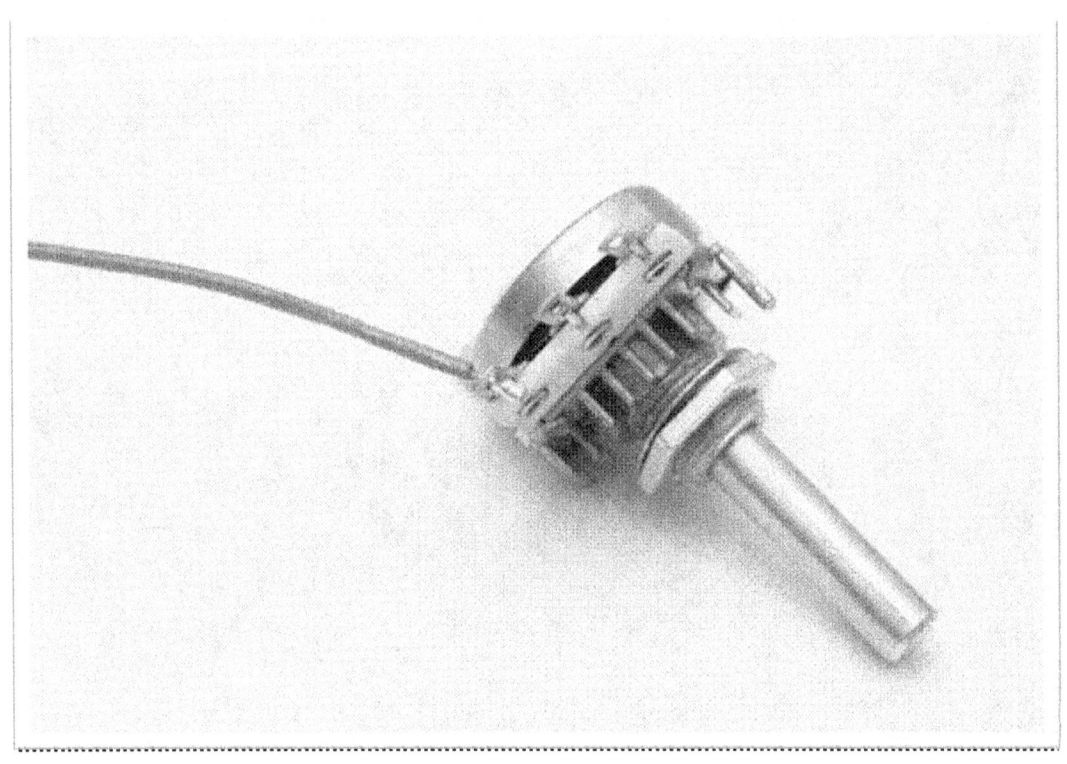

*^^ Then feed the tinned wire through the hole in the solder tag. Crop the wire with cutters if needed...*

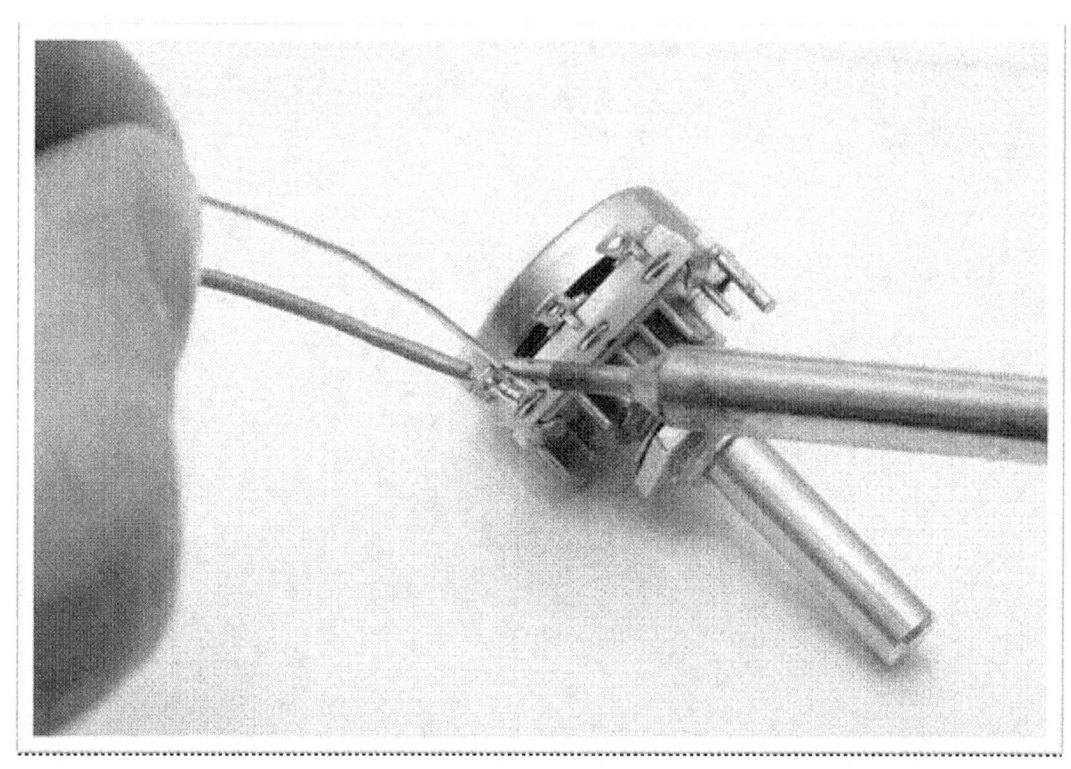

^^ ... *and apply the iron to heat the joint. As there's quite a lot of metal to heat, allow several seconds to heat it up and then simply solder the wire and solder tag together with a dab of solder wire.*

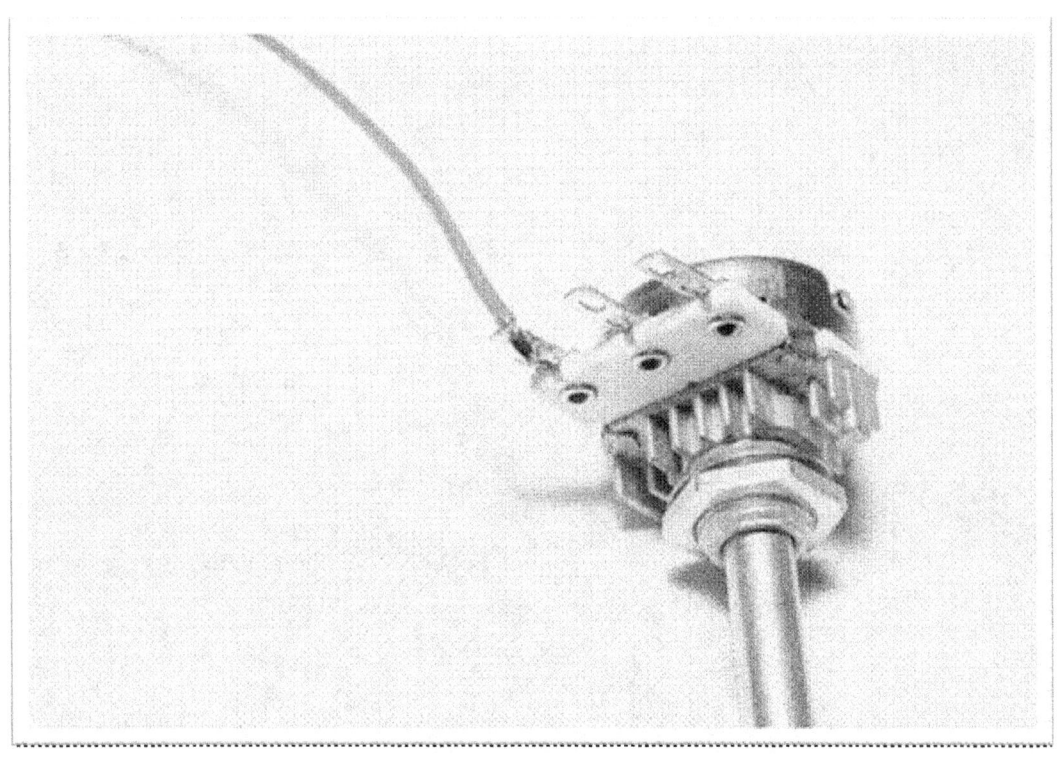

*^^ The result is a perfectly satisfactory solder joint.*

The second way is to loop the **untinned** wire through the tag a few turns and then solder it. This secures the wire during soldering, but it's messier to desolder if things go wrong (see later):

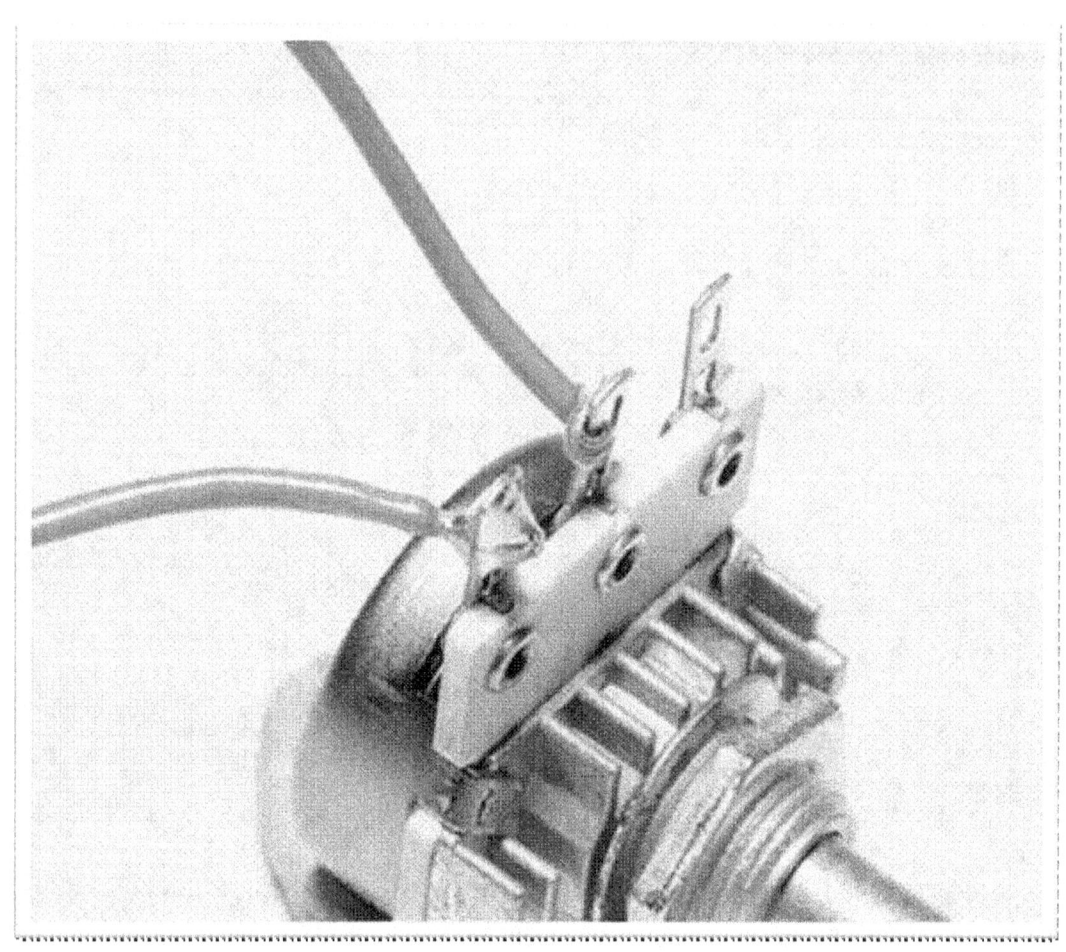

*^^ The centre wire (potentiometer wiper terminal) has been wrapped around to secure it before soldering.*

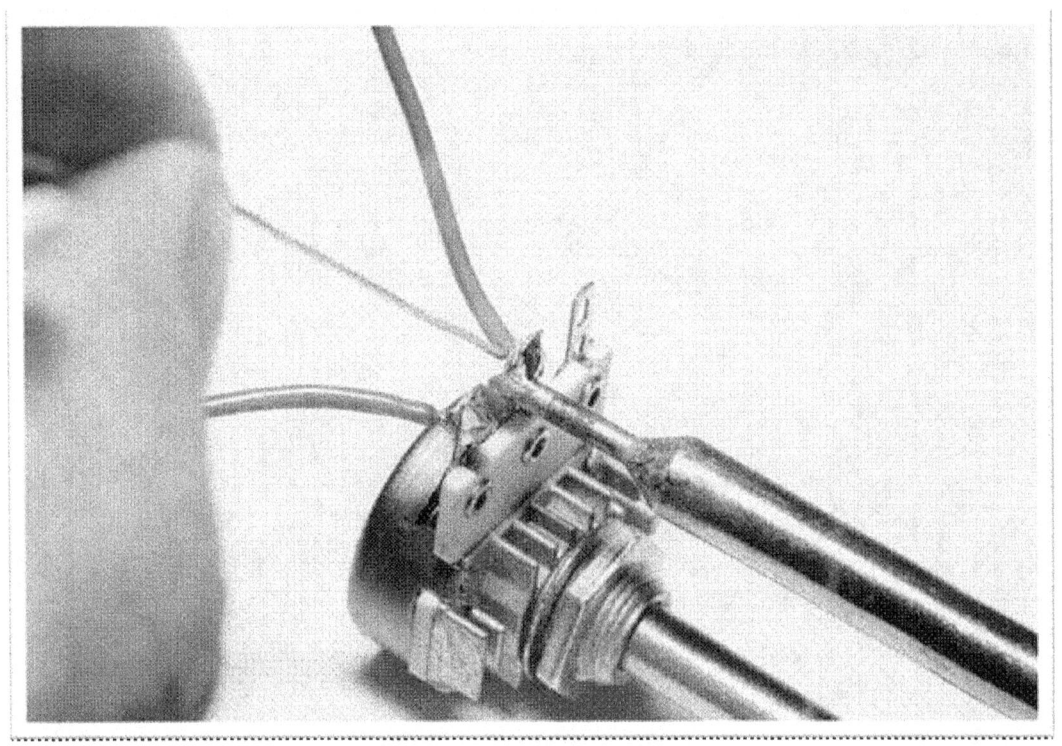

*^^ Solder the joint in just the same way – heat it up and dab on some solder wire.*

Much electronic equipment is connected or inter-wired this way and soldering everything together is a key stage in assembling any electronic project. I'll explain some refinements you can try, later.

## Wire joints

Soldering isn't always the best solution to some problems, for reasons that I explain in the next section called Fatigue and Breakage. However you can solder wires together to join them simply and cheaply, and next I show a classic wire joint of this kind, sometimes called a **Western Union** joint.

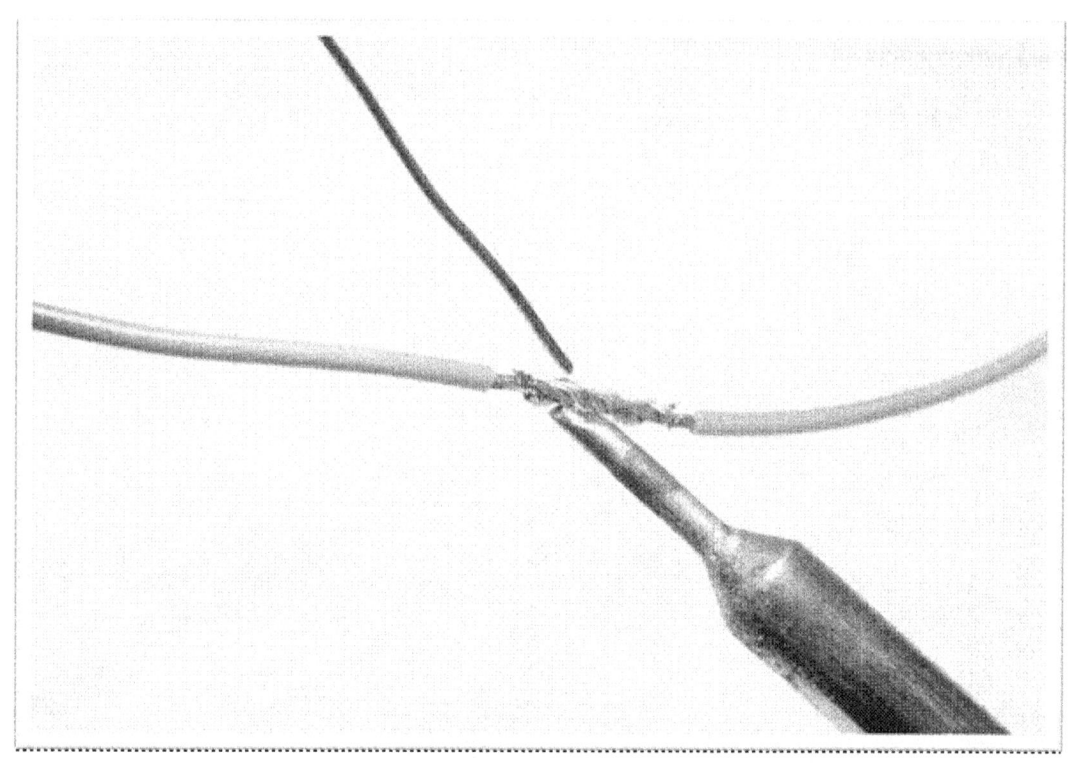

^^ *Soldering a Western Union wire joint.*

Strip the ends from each wire using wire strippers, and try not to cut any of the stranded cores or this will create a weakness. (Stray strands of wire can also be a hazard.) Then twist the wires securely together several times, ideally four or five. Apply a soldering iron to the exposed joint and dab on some solder until it flows fully over the joint, then remove the iron and allow the joint to cool.

The wire's plastic insulation may shrink back a little due to the heat, but try not to overheat it excessively. If you have the resources you can finish off by insulating the joint with some heatshrink tubing (slide it over a wire before you make the solder joint!). Otherwise use ordinary PVC tubing or insulation tape if you have any.

A solder tag can be soldered easily enough, and it's probably best to feed a stripped wire end through the tag and twist it over several times before soldering it, so it's held together when soldering.

*^^ A solder tag is soldered by wrapping a stripped wire end through the hole and round again, to secure it.*

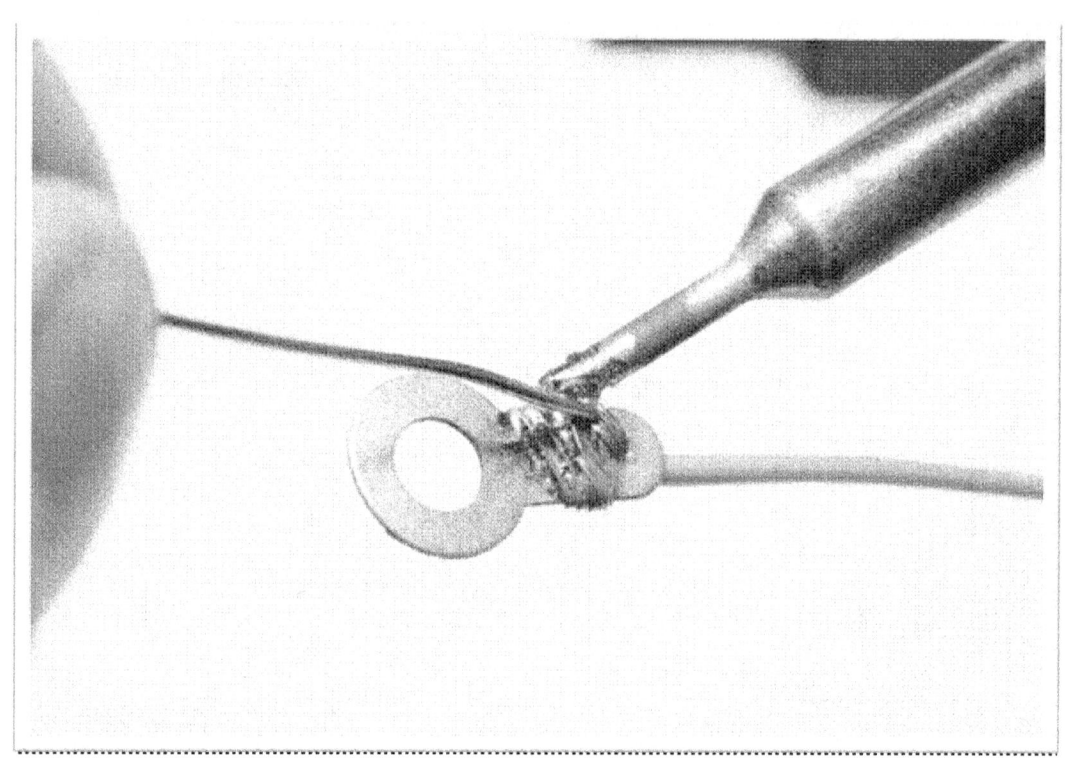

*^^ Apply plenty of heat and solder for a few seconds.*

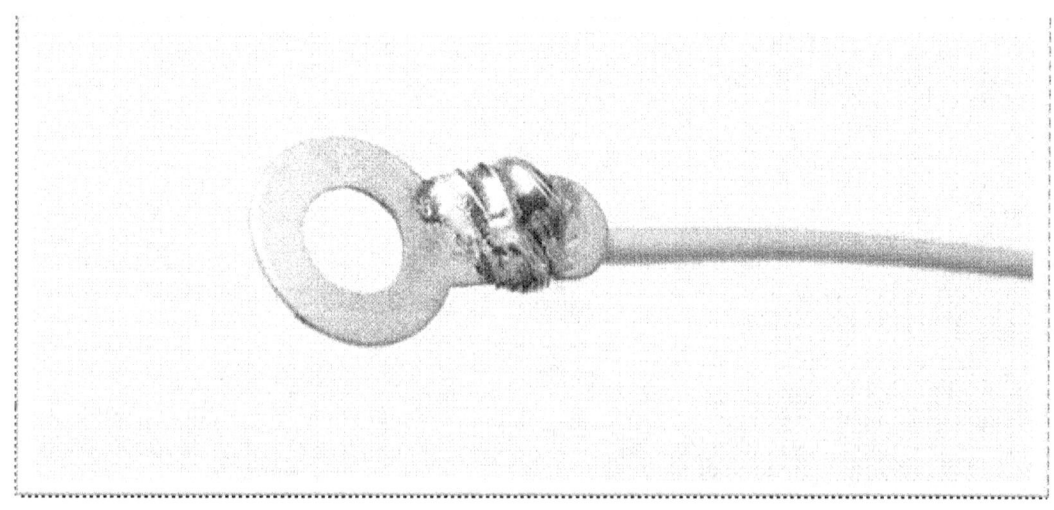

The next sequence show how the same basic principles of soldering wires are used to connect up a device such as a panel-mounting l.e.d.. A series resistor can be wrapped around and soldered directly to one of the l.e.d. solid wires, and then two multi-stranded connecting leads are soldered to the l.e.d. and resistor. It's best to use heatshrink tubing or PVC sleeve to insulate the solder joints and prevent short circuits afterwards.

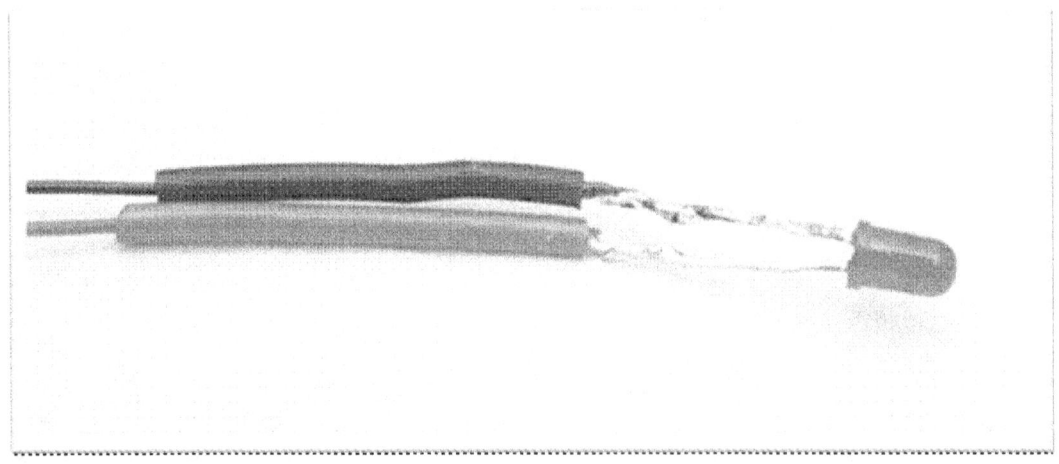

*^^ Multi-stranded wires can be soldered to the solid leads of an l.e.d.: insulate them afterwards. Different coloured wires identify the polarity of anode and cathode.*

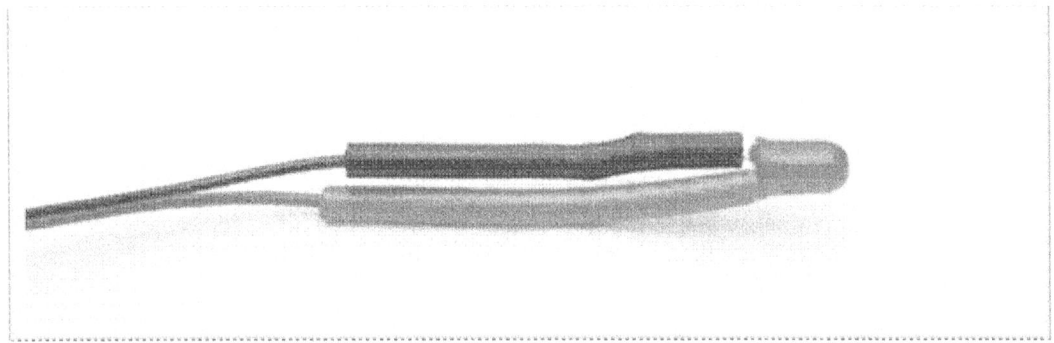

*^^ PVC sleeving or heatshrink is needed to prevent short circuits.*

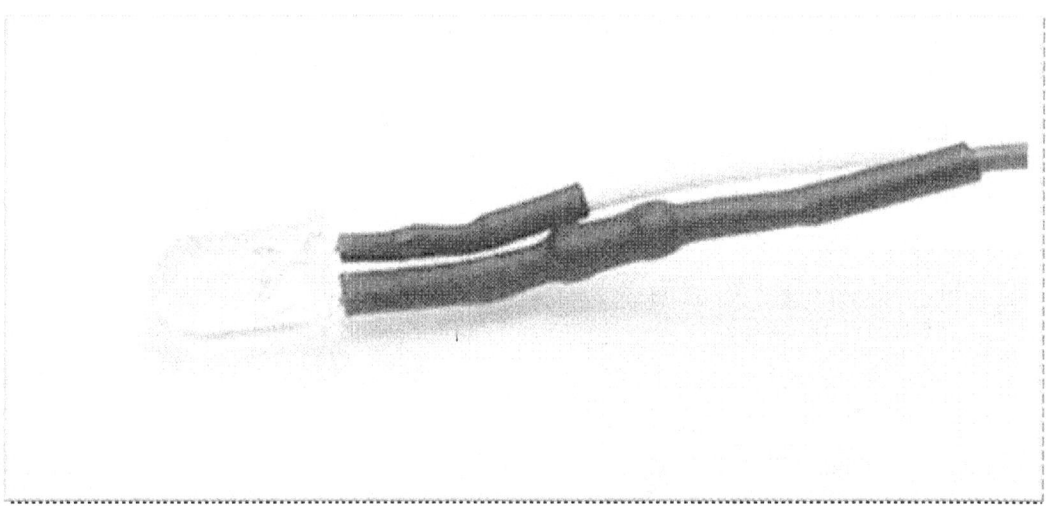

*^^ A commercial l.e.d.– the series resistor can be made out in the sleeving.*

Wires can also be soldered directly to printed circuit boards by stripping and tinning the ends, provided they fit through the holes in the p.c.b. It's a very common and cheap way of hooking up a board using "flying leads" and you'll see this all the time in consumer electronic equipment. For convenience so-called "solder pins" can be used, onto which flying leads may be soldered from the component-side instead.

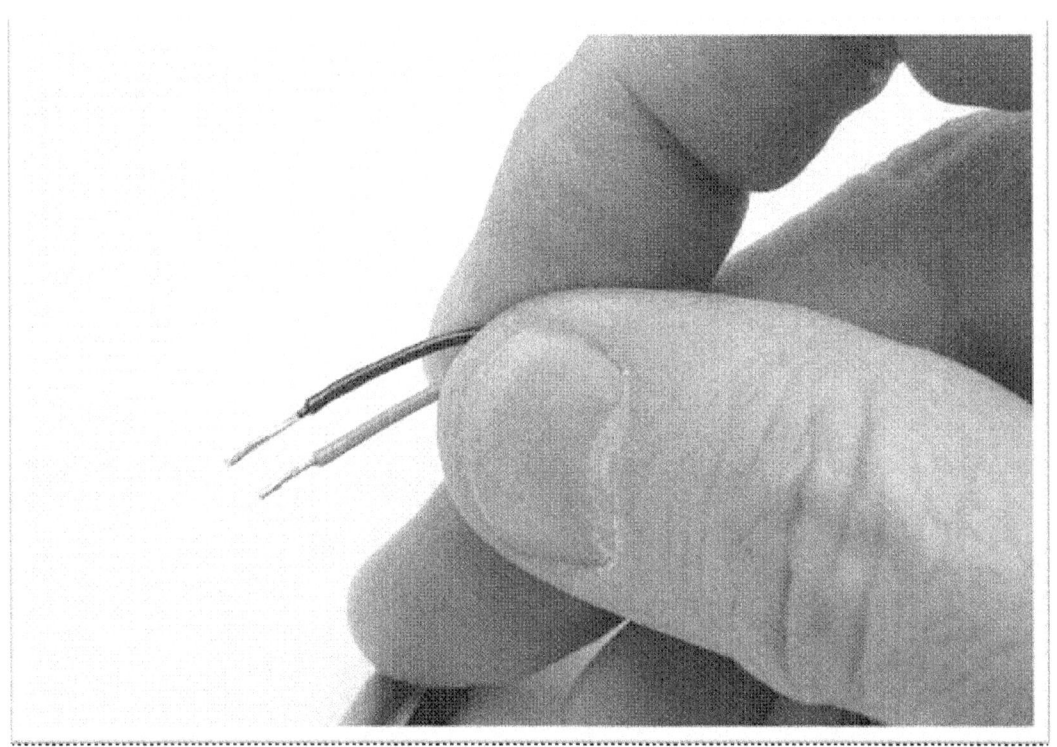

*^^ A set of l.e.d.s connecting wires, stripped, tinned and ready to be soldered to the p.c.b.*

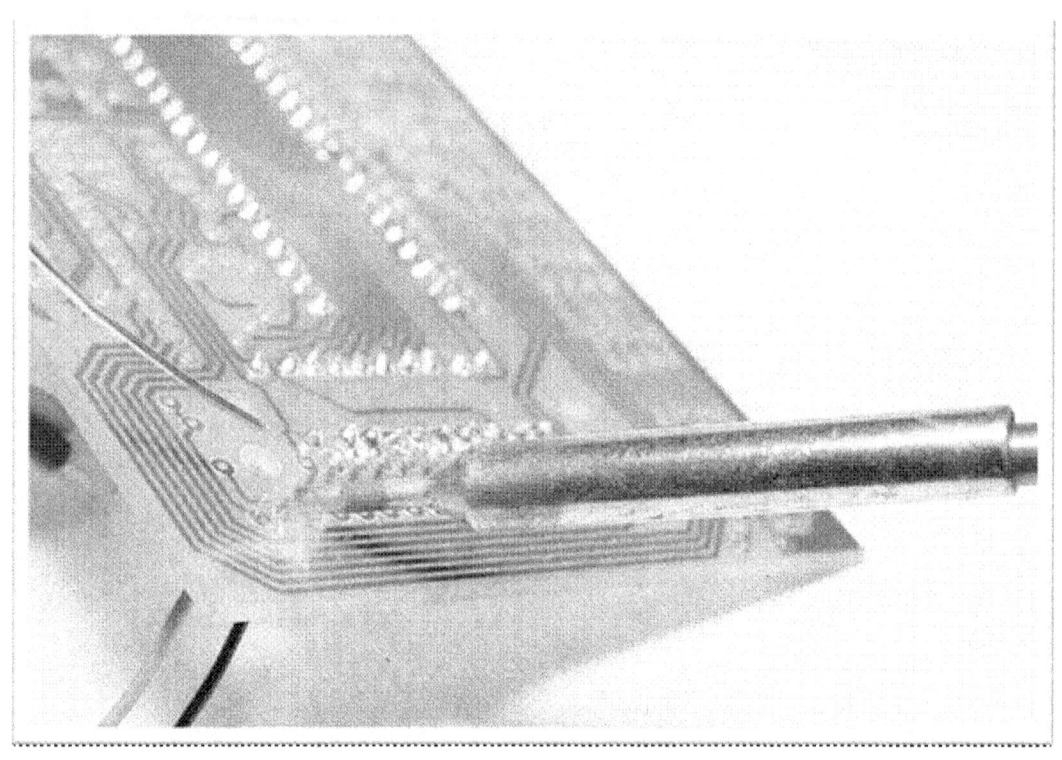

*^^ Solder the two "flying leads" to the copper pads in the usual way.*
*Hopefully the holes are big enough for the wires!*

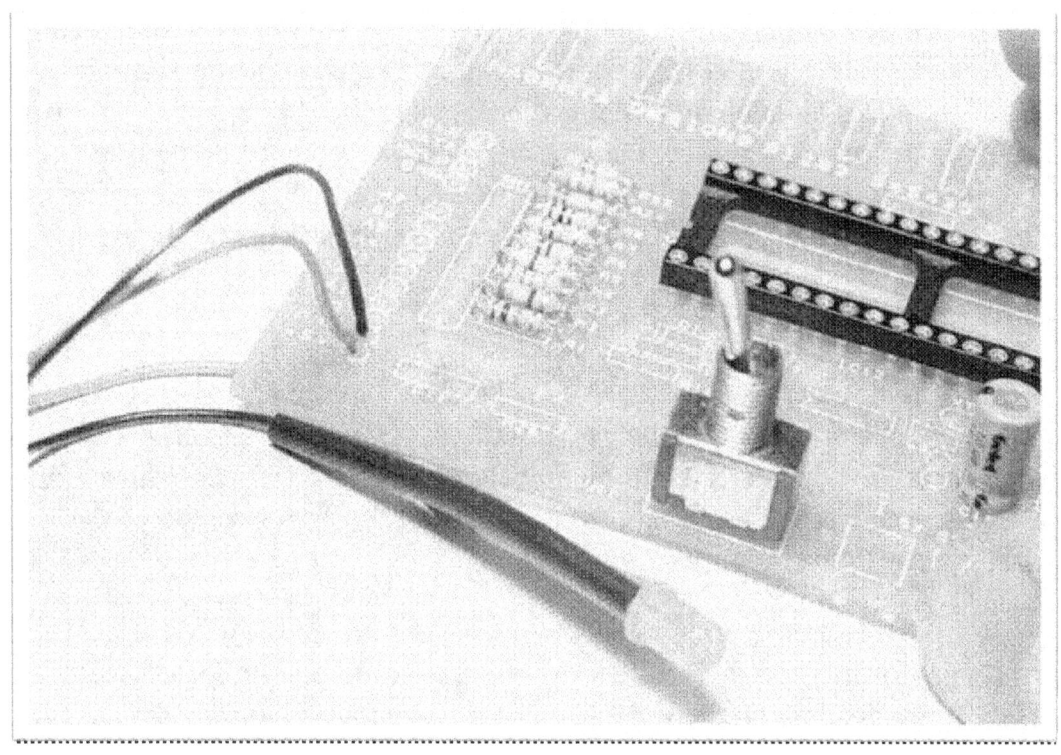

*∧∧ Connecting an l.e.d. to a circuit board with flying leads.*

Wires can also be attached to the underside (solder-side) of circuit boards, tacking them onto existing solder joints by re-melting them and absorbing the tinned wire end into it ("reflow soldering" – see later). This is a cheap and cheerful, semi-reliable way of rigging wires to a circuit board, and it's used all the time in imported consumer electronics.

## Tidying up

I'll leave it to readers to decide whether they should cut off excess wire before or after soldering. After the soldering is complete, I prefer to tidy up the joint by snipping any excess wire from the joint using a pair of "end cutters" shown earlier. These expensive hand tools have specially angled blades that snip the joint neatly down to the top of the solder joint. Ordinary side cutters will do fine.

It's worth taking time out to inspect the work closely, looking for any missing solder joints, whiskers of solder or swarf shorting out any solder pads, and all such potential problem areas should be dealt with prior to testing the board. Faultfinding goes beyond the scope of this guide, but it's true to say that almost always, any problems noticed after powering up the circuit are due to soldering faults or wrong components being used.

## Reflow technique

Another technique often used is "reflow" soldering. This is used to "tack" devices or wires together, especially if very small, sensitive or fiddly parts are involved. There might be no room to make a "proper" sturdy joint, or it might just not be necessary to have any mechanical strength in the joint, especially if tiny parts are used.

As an example, imagine a small temperature sensor (I used a transistor) for use in a thermometer project. It could be quite tricky to solder flying leads onto the sensor's leadouts, so a good approach is to tin both the flying leads and the sensor's leads, and then simply touch them together and re-melt the solder with the iron. There's no need to add any more solder, because the solder that's already there will re-melt and the joint will be made. Sometimes this is called a butt joint.

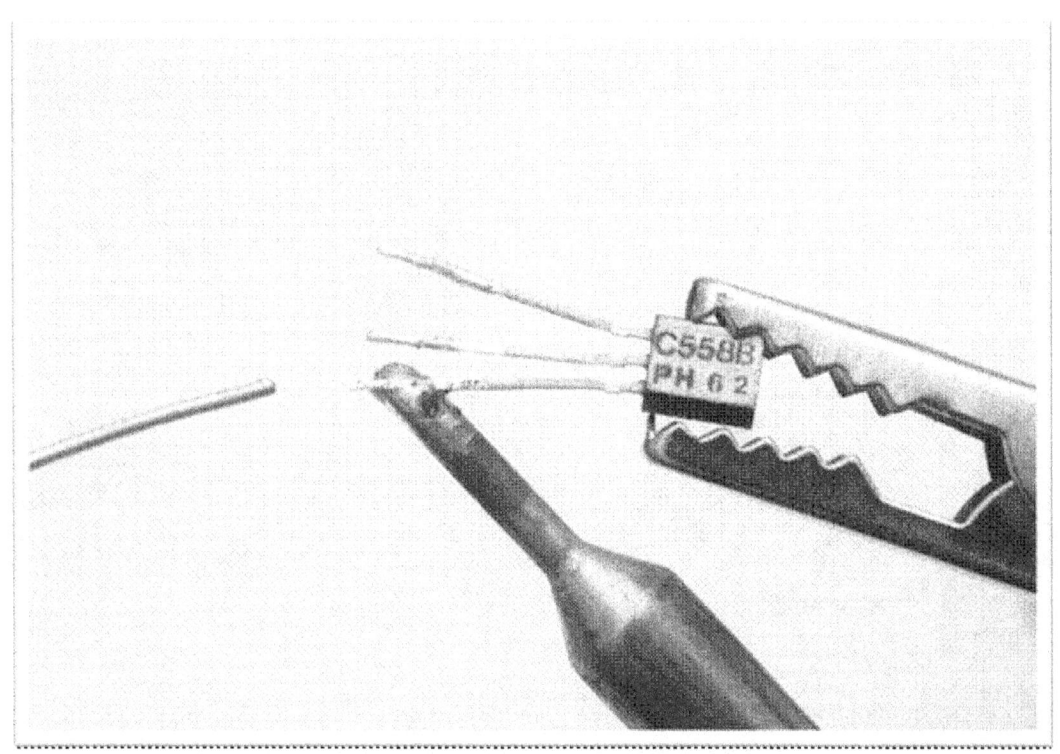

^^ *The leads of this sensor (transistor) have been tinned, ready for flying leads to be tacked onto them with a reflow method. A Helping Hands jig might help!*

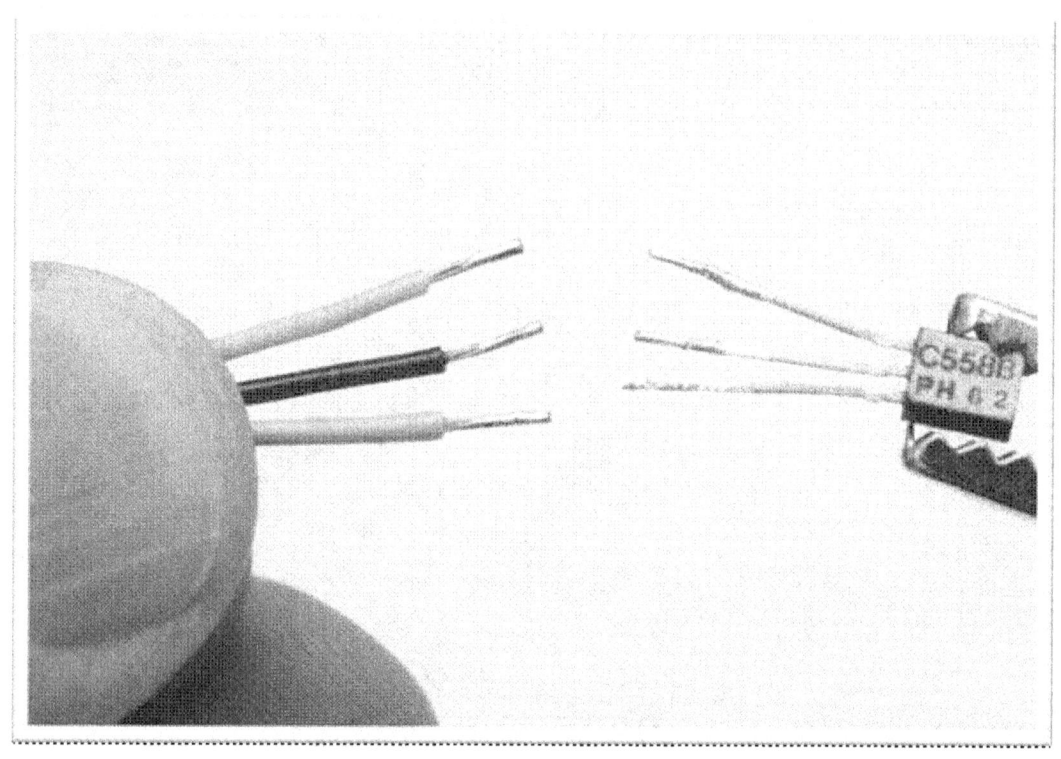

*^^ The three flying leads stripped, tinned and ready to be reflow-soldered onto the device.*

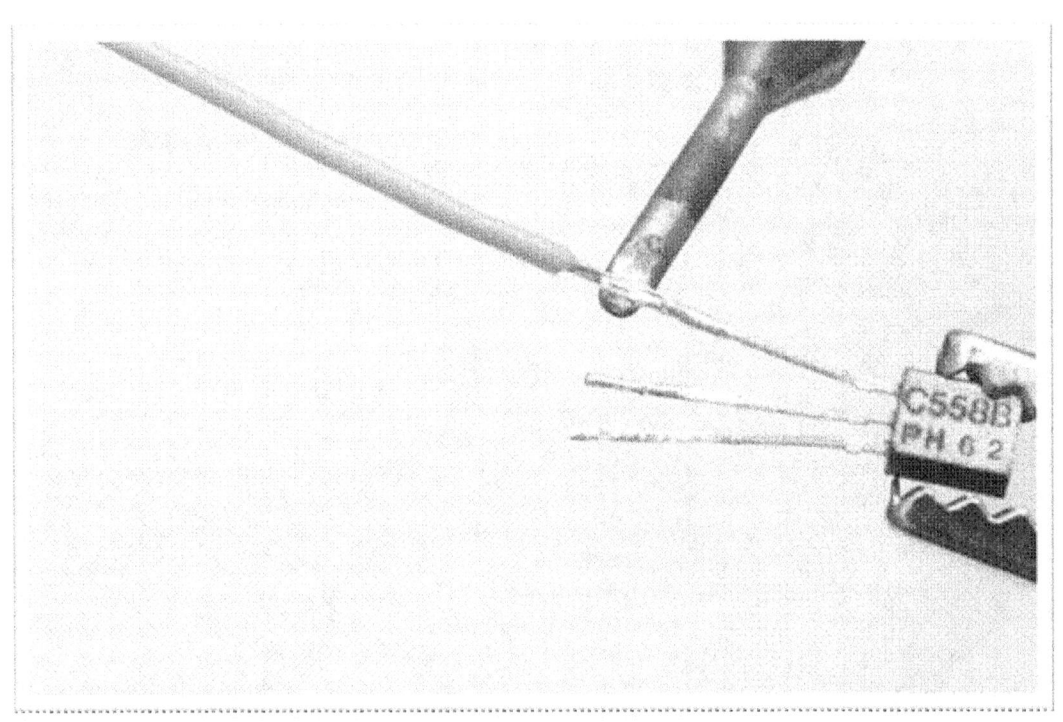

*^^ To reflow solder them, simply hold the two leads together while re-melting the solder with the iron.*

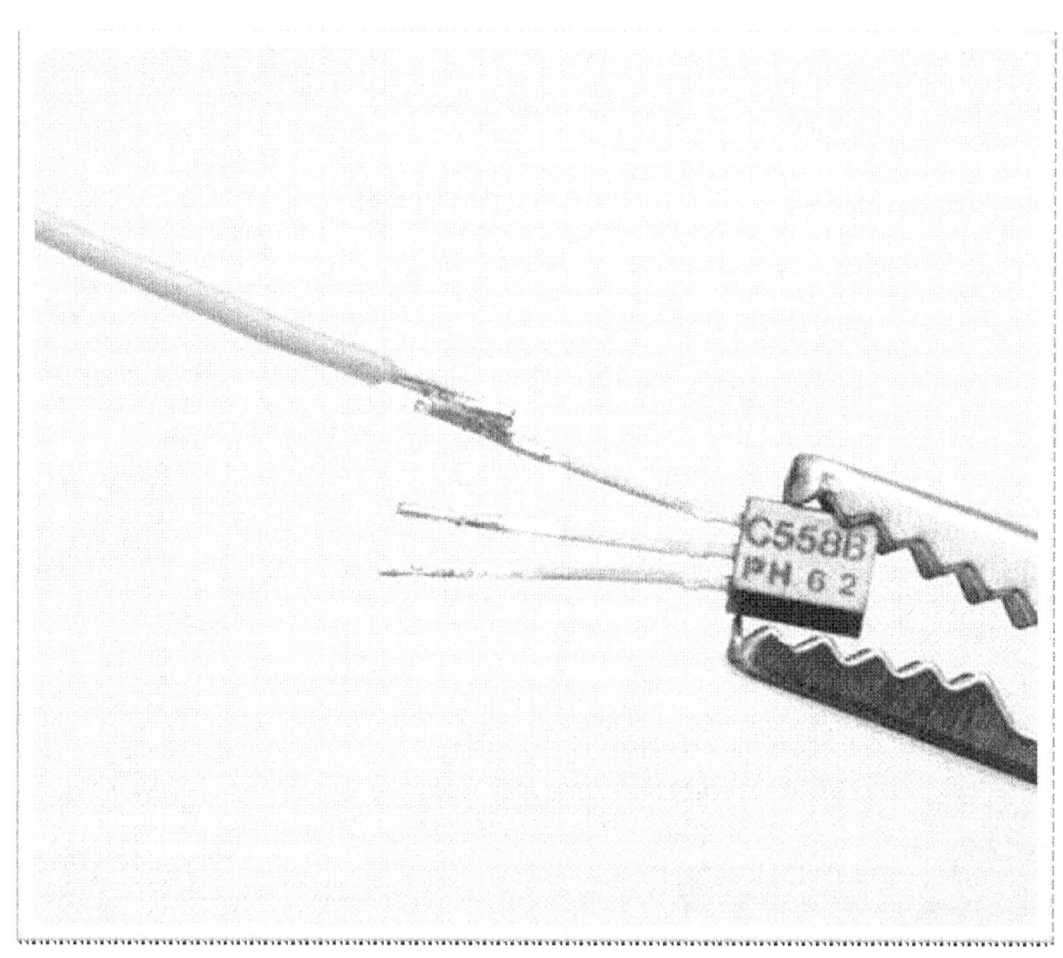

^^ *Remove the iron and let the solder cool. The wire is tacked on.*

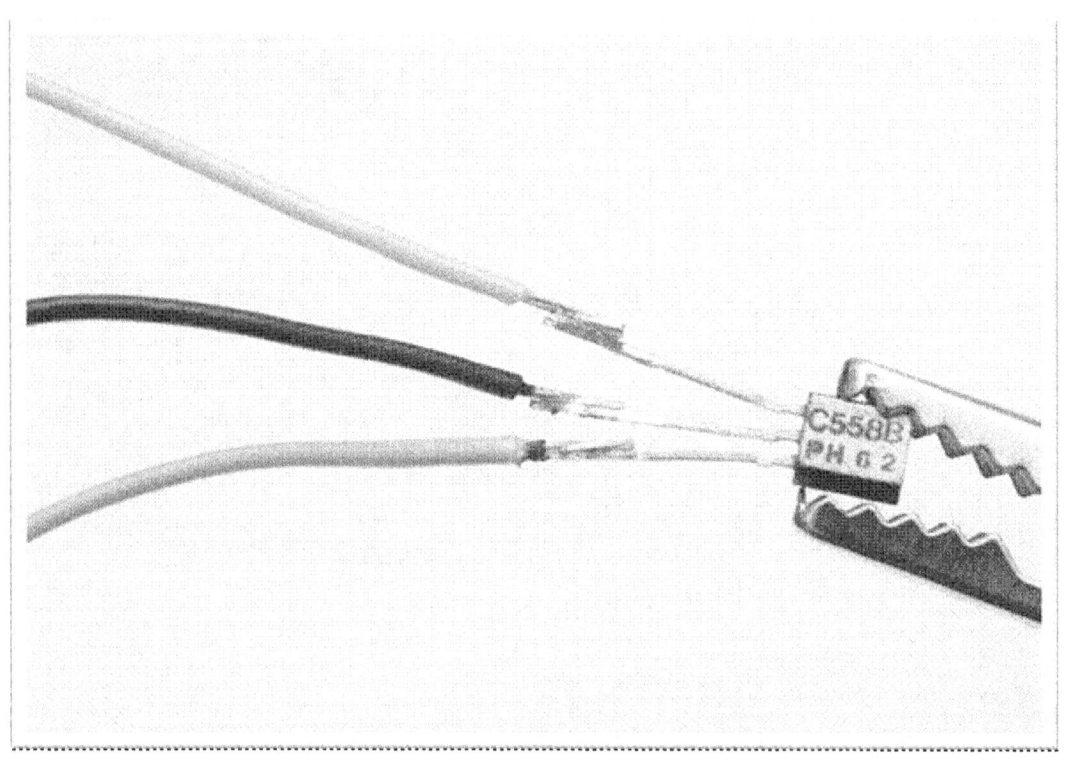

*^^ Repeat for the other leads. They just need insulating with sleeving, then job done!*

Depicted next is a typical "tag strip", an insulated panel with metal solder tags used for making sundry connections. (Entire TVs and radios used to be hand-built with them, in the early-mid 20$^{th}$ Century!) The principles of soldering are exactly the same, but more time is needed when applying the soldering iron because more metal is present, which needs more heat. You'll need more solder for bigger joints like these, so larger diameter solder makes a quicker job of it. Consider adding more flux (see earlier) to see if it helps.

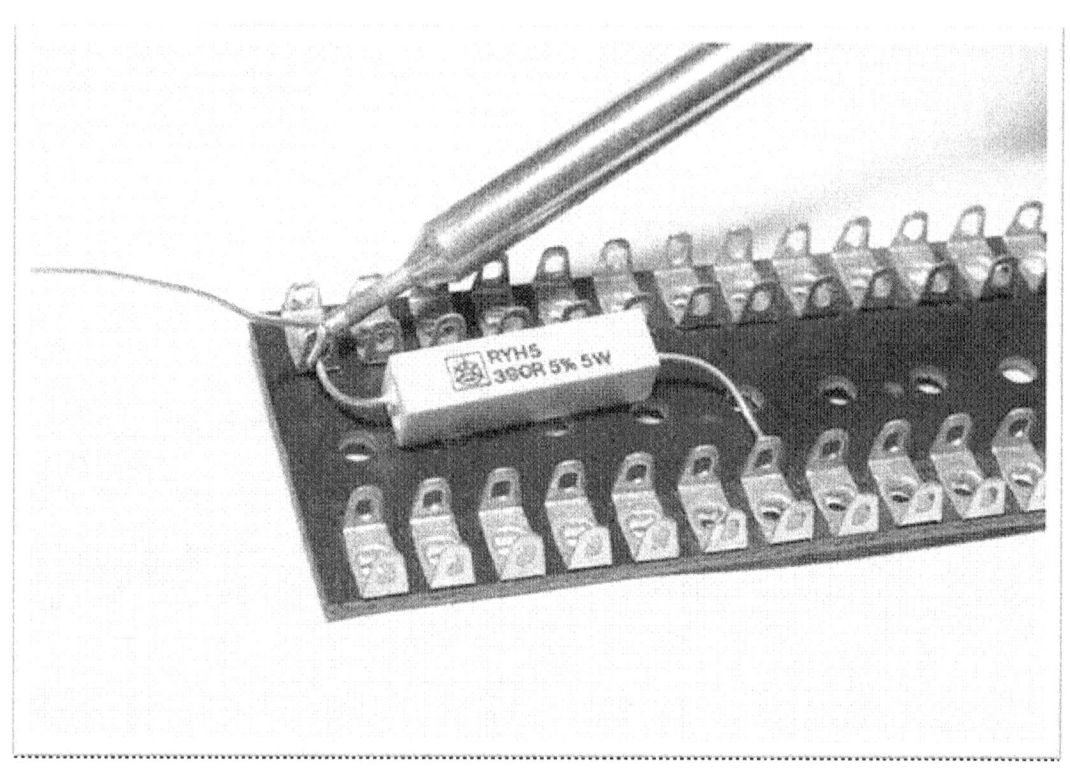

*^^ Wrap wires through the terminals and arrange everything neatly, then solder as normal.*

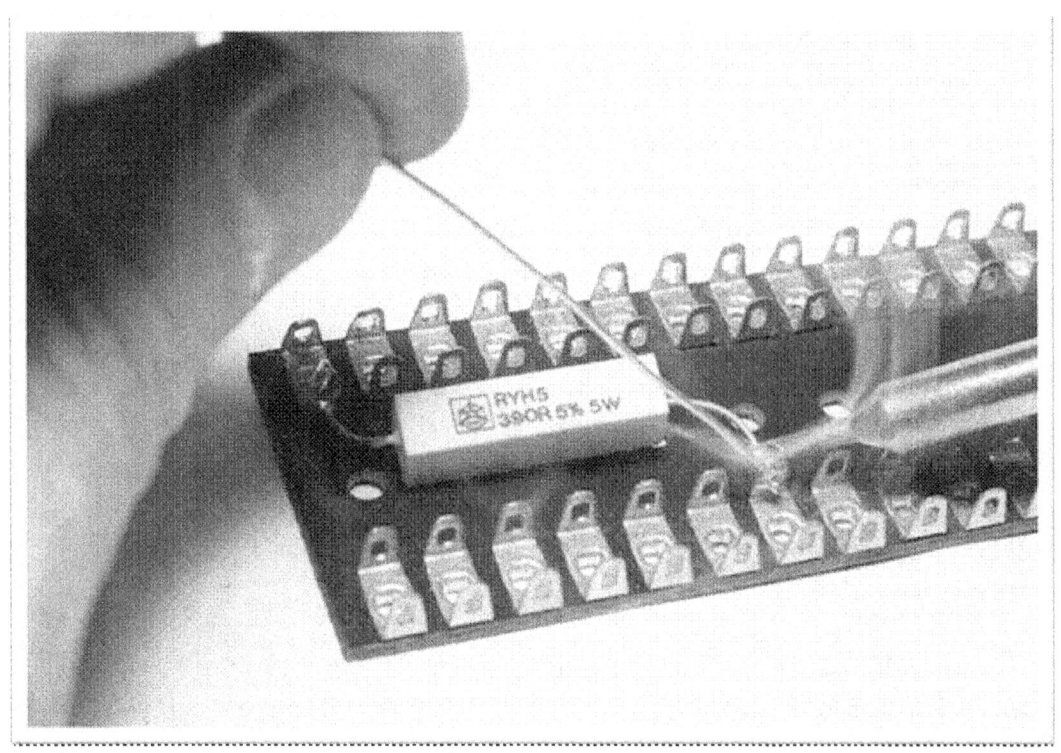

*^^ Don't be afraid to apply more heat with larger assemblies like these: they contain more metal than, say, an ordinary p.c.b so allow more time for the solder to flow over everything properly.*

## Fatigue & Breakage

Earlier I explained how multi-stranded hookup wire is flexible, which also makes it vibration proof – that's why a car's electrics are full of multi-stranded wiring, and test equipment probes use ultra-flexible multi-stranded wire for the same reason. Solid-core wire can be bent and will stay in shape (called "plate wiring") but if it's repeatedly bent or vibrated then it may eventually break somewhere due to fatigue.

The same is true of wires that have been soldered or tinned. No longer is the wire 100% multi-cored and flexible – instead it's been turned into a *single core wire* at the point where it's been soldered. This is potentially a weak spot and could eventually fracture due to fatigue, if subjected to continued vibration (e.g. in a car engine bay or in motorised equipment).

In a lot of equipment problems can be avoided by adding *strain reliefs* of some sort, to stop the wire being flexed where it's been soldered. Heatshrink tubing, or a dab of hot-melt glue, are ways of taking the pressure off the joint and ensuring soldered wires won't snap off due to vibration.

One of the reasons that higher-quality equipment and cars etc. use crimp terminals and connectors is that crimped (as opposed to soldered) connections retain all the flexibility of multi-core stranded wire from end to end, avoiding problems of wires breaking off. A *ferrule* is a very neat way of tidying wire ends and preventing stray strands of wire doing damage. Ferrules can be used to connect wires into screw terminal blocks etc. Simply clamp the terminal block down onto it and the screw will grip the ferrule.

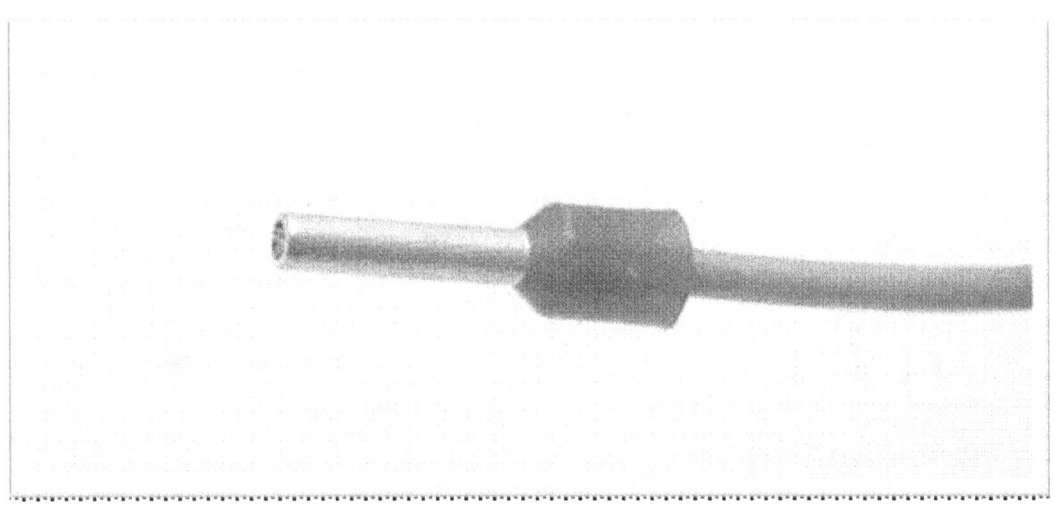

*^^ Instead of soldering them, wires can be terminated with ferrules prior to fitting to e.g. screw terminal blocks. This makes them vibration-proof and also avoids any problem of stray wires poking out.*

## Faults & Desoldering techniques

By putting into practice the guidelines in my *Basic Soldering Guide* practice, there's no reason at all why you should not obtain perfect results and eliminate any potential problems. Hopefully the information gives you plenty of guidance to tackle various soldering projects with confidence. There's no substitute for getting some hands-on experience though, so I'd repeat the advice to try assembling a simple high quality electronic kit or two, such as those produced by Velleman and see how you get on. Powering up your first project successfully is a great thrill as every electronics hobbyist knows.

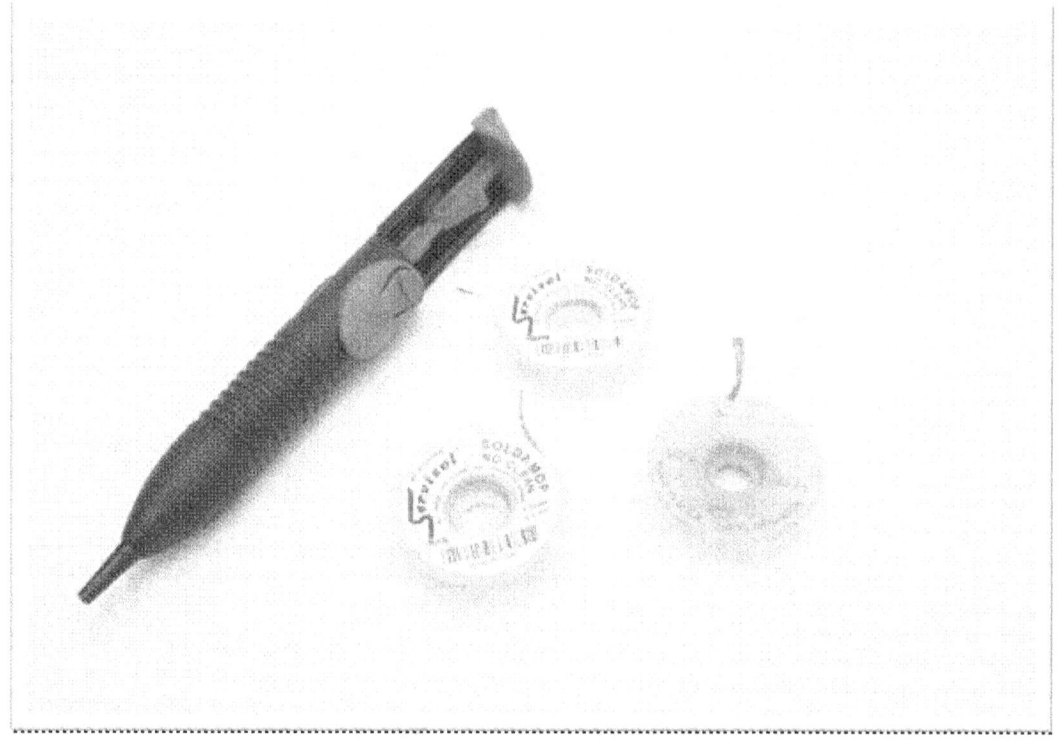

^^ *Popular desoldering products, a pump and various widths of desolder braid.*

Let's now look at reversing the soldering procedure – what to do if things go wrong, or maybe you have to repair a circuit by replacing a faulty component.

A solder joint which is badly made is likely to be electrically "noisy", unreliable and will probably worsen over time. Expansion and contraction of the joint due to heating and cooling can also throw up intermittent problems later down the line. The joint may look OK but underneath it may have a poor electrical connection, or could work initially and then cause the equipment to fail at a later date! These intermittent problems can be maddening to fix. TV repair technicians have an uncanny ability to go straight to a faulty solder joint because they see the same problem all the time, especially on equipment that has a "reputation".

A solder joint that's poorly formed is called a "dry joint" or "cold joint" or a "grey/ gray" joint. Usually it results from inadequate heating, dirt or grease preventing the solder from melting onto the parts properly, and is often noticeable because the solder tends not to wet the surface properly. Instead it forms beads or globules. Alternatively, if it seems to take a very long time for the solder to wet the joint, that's another sign of contamination and that the joint may be a dry one, or the material is incompatible anyway. A matt, crystalline appearance instead of a shiny joint points to inadequate heating: the solder cooled down far too quickly and didn't flow properly.

Whether you want to replace a faulty component or fix a dry or poor-quality solder joint, it's usually necessary to remove the troublesome solder and then re-solder it afresh. Naturally, there are tools and techniques that make the job easy. It's very bad practice to simply re-melt the joint and then lash out with the board, whiplash style, hoping that the molten solder will be flicked off the board.

The usual way of removing solder from a joint is to use a *desoldering pump*. These work like a small spring-loaded bicycle pump, only in reverse! A plunger is pressed down until it locks into position. It's released by pressing a button which sucks air back through a pointed nozzle, carrying any molten solder with it. It may take one or two attempts to clean up a joint this way, but a small desoldering pump is an invaluable tool especially for p.c.b. work and they are widely available now.

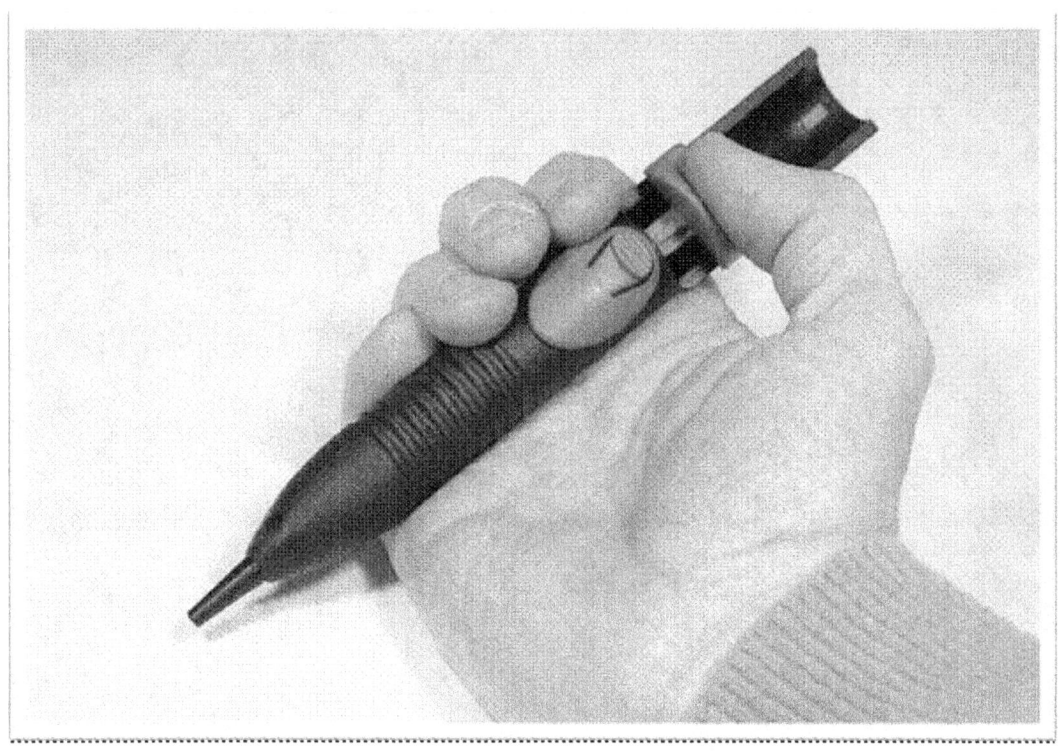

*^^ A hobby desoldering pump primed for use*

Desoldering pumps often have a heatproof P.T.F.E. nozzle which may need replacing occasionally. Each time the button is pressed, a plunger clears the nozzle but sometimes solder particles and swarf will be ejected in the process; when you prime the pump, point the nozzle into a small pot or old aerosol top to catch any debris. Remove the spout and clean out the pump from time to time.

*^^ Suck up molten solder using a desoldering pump*

With very stubborn joints where the last traces of molten solder just can't be shifted, it sometimes helps to actually *add* more solder and then desolder the whole lot again with a pump.

## Desolder Braid

An alternative to a pump is to use *desoldering braid* which arrives in small dispenser reels. It's a flux-impregnated fine copper braid which is applied to the molten joint, and the solder is then drawn up into the wick by capillary action. It's remarkably effective and for certain tasks, it can be more thorough than a pump. I recommend that a small reel is bought (start with 1.5mm width) for the toolbox, to tackle larger or difficult joints which would take several attempts with a pump.

*^^ Desolder braid is also handy and can sometimes be more effective than a desolder pump. It comes in various widths to suit the scale of work being tackled.*

To use desolder braid, press the end of the braid down onto the joint using the tip of an iron, and let the solder melt underneath: the braid will then absorb the solder. The braid becomes hot so beware of burns. Once the solder's solidified on the braid, cut it off and discard.

*^^ Desolder braid can also be used to remove excess solder, e.g. two i.c. pins shorted together.*

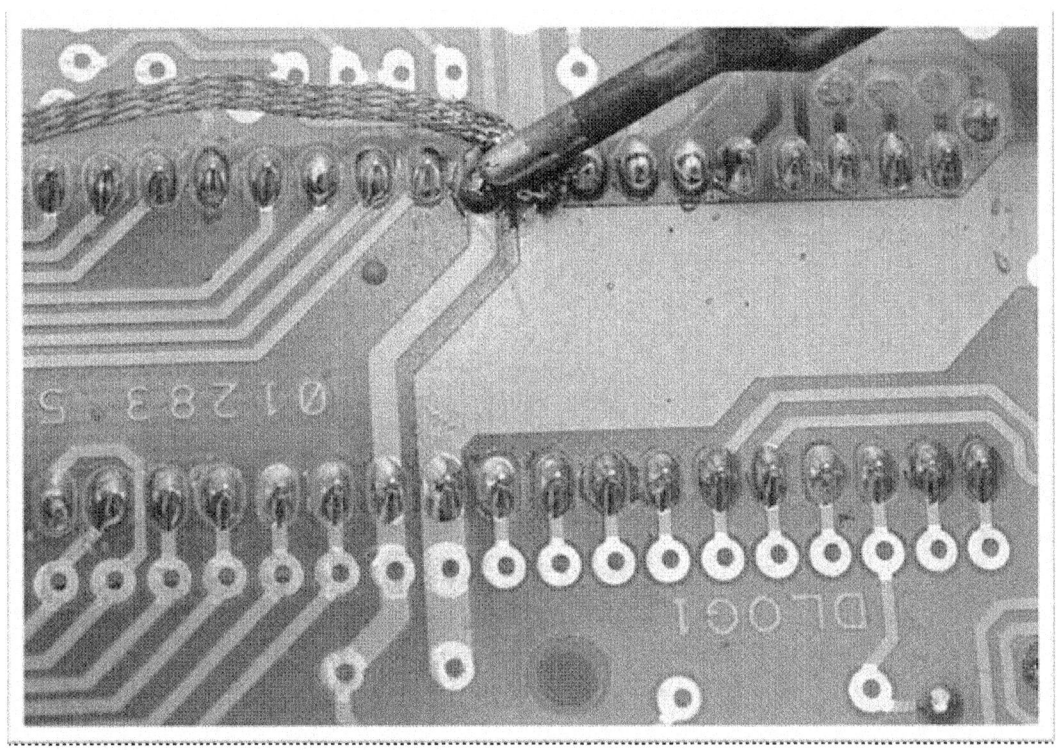

^^ *Press some desolder braid over the joint, then melt it with an iron to draw molten solder up the braid.*

*^^ Remove the braid immediately and don't drag "whiskers" of molten solder around. The excess solder is absorbed by the braid, which is snipped off ready for the next job.*

Be aware that you can damage a printed circuit board accidentally when removing the desolder braid if it's not removed *quickly enough*. The solder will soon solidify, which effectively solders the braid to the printed circuit board! A careless tug may yank copper tracks or pads away from the board, stuck to the braid. You can also drag solder "whiskers" onto neighbouring pads unless the braid is removed cleanly. Why not practice using desolder braid with an old circuit board?

Whichever desoldering method you use, care is needed to ensure that the boards and parts are not damaged by excessive heat. It's not that difficult to apply so much heat when desoldering that the adhesive holding the copper foil tracks on the p.c.b. eventually fails, causing the copper track to lift away – everyone's worst nightmare.

*^^ Typical copper track damage (centre) caused by overheating during soldering or desoldering. The track has lifted off, but you can try repairing it by adding extra wiring or SuperGlue it if the track isn't broken.*

If this should ever happen, remove the iron immediately and permit the area to cool (a freezer aerosol is valuable at such times). If you're lucky, you can maybe repair the lifted track using a droplet or two of Super Glue, or add "jumper wires" to bypass the damage.

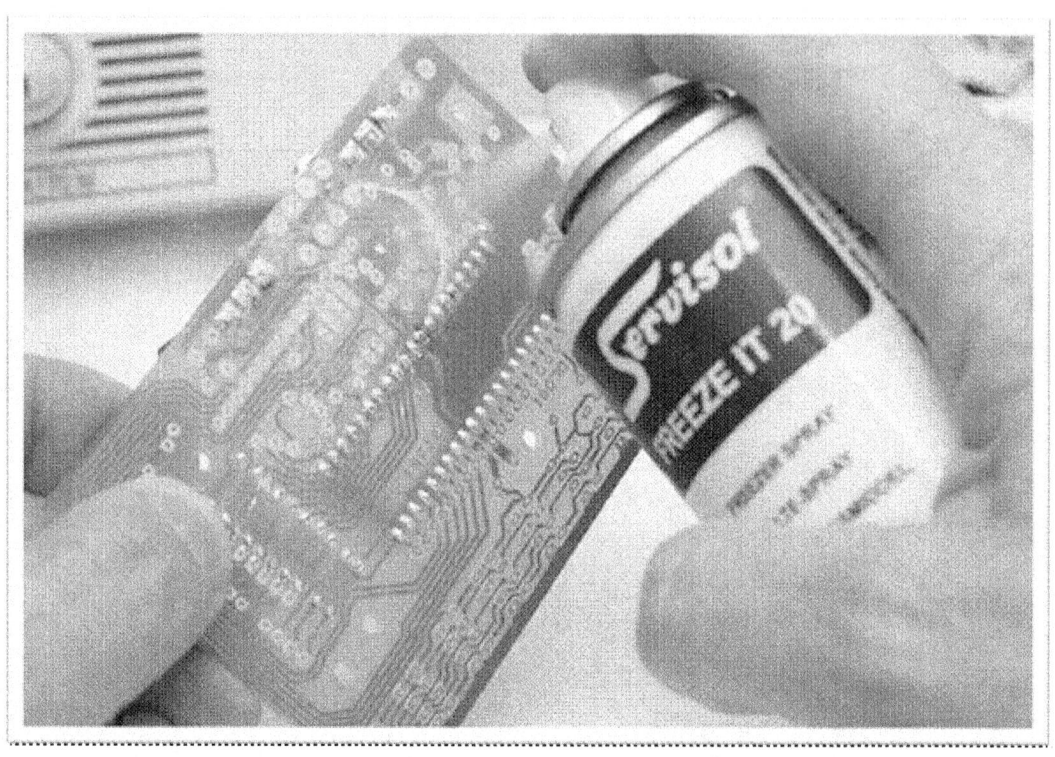

*^^ A Freezer aerosol can give rapid cooling where excess heat has been applied during soldering. Also used in circuit faultfinding to identify overheating parts.*

You now know everything you need to know about making the ideal solder joint, and desoldering it in case you need to make a repair. Just to remind you, a Quick Summary guide follows.

## Quick Summary Guide

To round off the Basic Soldering Guide, let's summarise how to make the perfect solder joint.

- Ensure materials to be soldered are compatible with tin/lead or lead-free solder.
- All parts must be clean and free from dirt and contaminants.
- Try to secure the workpiece firmly during soldering.
- Brand new soldering iron tips must be flooded with solder immediately, the first time they are used.
- Wipe the tip of the hot soldering iron on a damp cellulose sponge at frequent intervals. Then "tin" the iron tip by applying a small amount of solder.
- Aim to heat all parts of the joint with the iron for under a second or so, to bring them up to the same temperature.
- Continue heating and apply sufficient rosin-core tin/lead or lead-free solder to form a complete joint.
- It only takes a second at most, to solder the average p.c.b. joint. It should be smooth and shiny, and through-hole joints should be slightly convex in shape.
- Remove the iron and return it safely to its stand.
- Do not move parts until the solder has cooled.
- Tin the soldering iron tip and clean it well, when switching it off, ready for next time.
- Consider using e.g. electronics flux dispenser pens or Colophony (rosin) to help with difficult joints.

Sometimes solder joints don't go quite to plan, and sooner or later everyone is faced with the need to problem-solve or troubleshoot, so a simple Troubleshooting Guide follows next.

# Troubleshooting Guide

This troubleshooting guide may help fix common problems encountered with troublesome solder joints.

| SYMPTOMS | LIKELY CAUSES | REMEDY |
|---|---|---|
| Solder won't "take" (wet) and won't flow properly over the joint -- molten solder forms beads or "ball bearings" instead of flowing properly. | Grease or contaminants present; Material may not be suitable for soldering with standard lead/tin or lead-free solder, e.g chromium. | Treat contaminated parts with abrasive cleaners etc. as required to expose base metal. Some metals can't be soldered with electronics-grade solder. |
| Solder doesn't melt or flow very well -- the joint is crystalline or grainy-looking - a grey or dry joint. | Joint has been moved before being allowed to cool naturally, or: Joint was not heated adequately. Too large a joint – too much metal present – and/ or the soldering iron temperature or power rating are too low. | Desolder and remake. Apply heat for a longer period, or use a higher power soldering iron, or check the temperature setting and raise it if possible. |
| Solder joint forms | Probably overheated, | It is usually best |

| | | |
|---|---|---|
| a "spike" and applying the iron again makes it even worse! | burning away the flux. The iron, when removed, would cause the solder to stand up in a spike. | to desolder and remake the joint freshly again. |
| The copper foil of my p.c.b. has lifted off the circuit board! | Excessive use of heat has damaged the adhesive. Provided the track hasn't broken, it may be repairable. | You can sometimes repair it with Super Glue, or re-wire the board with jumper wires. |
| Brown varnish-like deposits are left behind after I finish soldering. | These are the remains of rosin flux and are nothing to worry about. | It can be removed with PCB cleaners or some solvents, if you want to tidy up the board and inspect your work. |

## Possible Hazards and simple First Aid

It's extremely rare that soldering iron operators receive any burns or other injuries from the use of hot soldering irons. Soldering is perfectly safe provided that common sense precautions are taken during the soldering operation. Here are some of them:

- Components are very hot after soldering, so let them cool before handling them to avoid skin burns.
- Beware of splashes of molten solder caused by careless handling of a hot soldering iron.
- Beware of energised components (capacitors, batteries etc.) being shorted by molten solder and ejecting solder splashes due to arcing.
- Always park a hot iron safely on a stand in between use — never hang it vertically next to the bench.
- Keep a hot soldering iron away from its mains cable (silicone cables reduce the risk of accidental damage).
- Beware of wire offcuts flying off (danger to eyesight) when snipping wires to length before or after soldering.
- Avoid inhalation of solder and flux fumes as this can irritate the respiratory tracts, especially in sensitive cases (e.g. asthma).

Should you receive a more serious skin burn which requires attention, then:

- **Cool the affected area immediately.** Use plenty of cooler running water – but avoid ice cubes etc. as they can cause nerve damage after a time or inhibit the flow of blood to the affected area.
- Remove any **objects** which may be constrictive, before any swelling starts (rings, watches, bracelets).

- Do not prick blisters nor apply ointments, salves or lotions at this stage.
- Local pain relief for small burns can be obtained by spraying *Burneze* aerosol onto unbroken skin.
- Seek medical attention for more serious burns.

Eyesight problems are exceptionally rare, e.g. pieces of wire offcuts or solder splashes lodging in the eye area, and should be treated by a qualified first-aider or A&E. The best you can do is bathe the affected area with e.g. a first-aid eyewash bottle or fresh water. Then seek professional medical help straight away.